WITHDRAWN
D0204972

RADIO

Recent Titles in
Greenwood Technographies

Sound Recording: The Life Story of a Technology
David L. Morton Jr.

Rockets and Missiles: The Life Story of a Technology
A. Bowdoin Van Riper

Firearms: The Life Story of a Technology
Roger Pauly

Cars and Culture: The Life Story of a Technology
Rudi Volti

Electronics: The Life Story of a Technology
David L. Morton Jr. and Joseph Gabriel

The Railroad: The Life Story of a Technology
H. Roger Grant

Computers: The Life Story of a Technology
Eric G. Swedin and David L. Ferro

The Book: The Life Story of a Technology
Nicole Howard

RADIO

THE LIFE STORY OF A TECHNOLOGY

Brian Regal

GREENWOOD TECHNOGRAPHIES

GREENWOOD PRESS
Westport, Connecticut • London

Library of Congress Cataloging-in-Publication Data

Regal, Brian.
Radio : the life story of a technology / Brian Regal.
p. cm.—(Greenwood technographies, ISSN 1549–7321)
Includes bibliographical references and index.
ISBN 0–313–33167–7 (alk. paper)
1. Radio—History. I. Title.
TK6547.R44 2005
621.384'09—dc22 2005017403

British Library Cataloguing in Publication Data is available.

Library of Congress Catalog Card Number: 2005017403
ISBN: 0–313–33167–7
ISSN: 1549–7321

First published in 2005

Greenwood Press, 88 Post Road West, Westport, CT 06881
An imprint of Greenwood Publishing Group, Inc.
www.greenwood.com

Printed in the United States of America

The paper used in this book complies with the Permanent Paper Standard issued by the National Information Standards Organization (Z39.48–1984).

10 9 8 7 6 5 4 3 2 1

Contents

Series Foreword

In today's world, technology plays an integral role in the daily life of people of all ages. It affects where we live, how we work, how we interact with each other, and what we aspire to accomplish. To help students and the general public better understand how technology and society interact, Greenwood has developed *Greenwood Technographies*, a series of short, accessible books that trace the histories of these technologies while documenting *how* these technologies have become so vital to our lives.

Each volume of the *Greenwood Technographies* series tells the biography or "life story" of a particularly important technology. Each life story traces the technology, from its "ancestors" (or antecedent technologies), through its early years (either its invention or development) and rise in prominence, to its final decline, obsolescence, or ubiquity. Just as a good biography combines an analysis of an individual's personal life with a description of the subject's impact on the broader world, each volume in the *Greenwood Technographies* series combines a discussion of technical developments with a description of the technology's effect on the broader fabric of society and culture—and vice versa. The technologies covered in the series run the gamut from those that have been around for centuries—firearms and the printed book, for example—to recent inventions that have rapidly taken over the modern world, such as electronics and the computer.

While the emphasis is on a factual discussion of the development of the technology, these books are also fun to read. The history of technology is full of fascinating tales that both entertain and illuminate. The authors—all experts in their fields—make the life story of technology come alive, while also providing readers with a profound understanding of the relationship of science, technology, and society.

Acknowledgments

I would like to thank Kevin Downing and everyone at Greenwood Press. Also, thanks to Seyed Akhavi and Paul Levinson for reading over parts of the manuscript (though I take responsibility for any factual errors).

Introduction

When the *RMS Titanic* went to its icy grave in 1912, it took with it more than just most of its crew and passengers. It also took with it the notion that a modern society could survive without a speedy mode of mass communication. The working definition of radio used here is that it is a mechanical device that operates by transmitting and receiving sound in the form of electromagnetic waves. Sound is presented to a transmitter in the form of voice or music. This sound is then transformed into an electrical impulse, which is amplified and then transmitted on a specific wavelength and at a specific frequency into the atmosphere as an electromagnetic wave. A receiving station adjusts itself to the same wavelength and frequency as the signal. The receiver can then convert the electromagnetic wave back into sound, which is amplified and sent out through the receiver's speakers to be heard. This definition is a bit incomplete because it covers only the mechanical aspect. The technology of radio evolved into something less tangible than just a box out of which sound came. It became a cultural force unlike few others. Along with print, television, and film, radio transformed human society in ways never anticipated. Radio took on a life of its own by ceasing to be a hard technology only and becoming part of the fabric of life. The unique relationship between radio and listener was crucial to its success. Despite the public nature of radio and the fact that millions might be listening at once, to a specific program, news flash, or religious sermon, there was still a one-on-one feeling. The voice coming out of the box was directed to a specific listener as if it was an intimate conversation. Whether it was a talk show host, whacky deejay, commercial pitchman, president or priest talking, they were talking to you. The mind was forced to work and think and ponder. Radio became more than just a useful technology. That is the reason this book is organized the way it is.

This is a biography of radio from two points of view: as machine and as media. The first half of the book will tell the history of the design and development of radio communications. It will cover its beginnings as an improvement for sending telegraph messages to its use as a voice transmission device. The second half of the book will deal with media history—the radio's impact on society and its place as cultural icon. Each half will consist of three chapters. There is also an historical chronology, an index, a glossary, and bibliography. The opening chapter will tell the story of the early discoveries that paved the way for radio. It will begin with sixteenth-century experiments into the nature of electricity and continue through the nineteenth-century and Faraday's dynamo, Maxwell's electromagnetic theory, and the invention of the telegraph and telephone. In addition to the history of theories and inventors, the scientific principles by which they operate will be discussed as well. Chapter 2 will include a series of biographies of the central figures from the developmental days of radio like Guglielmo Marconi, Nikola Tesla, Lee De Forest, Edwin Armstrong, and others. Chapter 3 will take the technological advancement of radio up to the present. The radio, because of its early mechanical limitations, was a private mode of communication. It began as a corporate and military device, which was also used by a few pioneering amateurs in the 1920s who began listening to broadcasts and then setting up their own transmission stations. Chapter 4 will deal with the solitary nature of the radio that was pushed into the public sphere almost against its will. Chapter 5 will cover the Great Depression onward when the radio came into its own as a mass communication media and went from a toy for hobbyists to a means of survival. The chapter will discuss radio as a public forum from the uplifting messages of FDR's Fireside Chats to the darker side of Father Coughlin (the "Radio Priest"), and use of the radio by the Nazis. After World War II, music became the prime mover of radio, allowing it to become an agent of social change in another way. During this period the modern radio personality was born. Intellectuals turned their attention to what radio was and what it was doing. Philosophers like Marshall McLuhan began to ask, "Is the media the message?" In Chapter 6 postwar radio became more an entertainment than a news provider, but by the end of the twentieth century returned to its interactive roots with the birth of the talk show and call-in programs.

People readily accepted the idea of listening to disembodied voices coming out of a box. Harcourt Index.

Science and technology have always been seen as related but separate issues. Science is the search for knowledge about the nature of the universe while technology is a practical application of knowledge to do something. However, by the end of the twentieth century something called techno-science emerged. This was the blurring of the traditional separation of the two. Where one ends and the other begins became less clear. As Ruth Schwartz Cowan said in her *A Social History of American Technology*, "the transformative power of twentieth century technology is derived from the investigative and creative power of techno science" (Cowan 1997, 222). This idea can certainly be applied to radio. Radio can be seen as a model of the conflict between science as provider of serious knowledge and advancement, and science as purely entertainment. Once radio reached its mature mechanical form, it ceased to be a machine and became an agent of social change used by various factions to put forward their agendas and beliefs and to rally the faithful to their causes. As a media of mass communication, the term radio ceased to describe a novel machine and became "radio," a larger issue. It ceased to be a technology at all in the eyes of its nontechnological users and became an abstract idea.

The biography of radio is complex. As it became an integral part of human society, the story became more than just one of nuts and bolts and vacuum tubes. It must be told as part of society not separate from it, and must encompass, in addition to the history of science and technology, the history of business and the corporate world along with the history of society and culture. As a result, the story must interweave material from seemingly diverse areas. This book is part of a series on the biography of technology. As with any biography, every detail of the subject's life cannot be covered in detail. This work is meant for those with little or no background in the area. The material here was chosen to give an overview of what the progress of radio entailed and to tell an entertaining and informative story. It should not be viewed as complete or definitive, but rather as a concise history designed to get the reader started. The story is not just about radio itself, but about the people who used it, abused it, listened to it, and tried to take control of it. The story is full of strange characters, deep thinkers, visionary mystics, mechanics, hyperactive minds, ambitious souls, power hungry demagogues, and utopian humanists. They all sought to make radio into what they thought it should be.

It will be useful at this time to briefly mention how to read a work like this. History, the writing about the past, can be a tricky thing. The past is everything which came before now. It is a collection of elements floating in space. The past becomes history when humans take those elements and organize them as a way of understanding them. A history is a story and stories have a beginning, middle, and an end. This type of structure suggests a closed system and connections between things that may not exist. The past is not a closed system. It is endlessly complicated, intertwined, abstruse, and often confusing. History is not a nice neat package with only one or two possible outcomes or interpretations. Many events, discoveries, and people will be discussed in this history of radio. Some inventions came before others, some ideas developed before others. It is easy to think that one thing led directly to another, but this is misleading. History is chaotic and does not always lead from point A to point B in nice straight lines. What people often see as connections of historical events are commonly only seen after the fact. Humans like patterns and predetermined outcomes. It is comforting to think that things were "meant to be" or are the "obvious result of" things that went before. What the reader must understand is that while the facts (the past) remain relatively stable, interpretation of those facts (history) often changes. Disagreement over historical interpretation should not be seen as a problem. Open debate over differing points of view leads to new and deeper insights about a subject.

Timeline

600 B.C.E.	Greek philosophers observe an effect after rubbing amber on animal fur. The Greek word for amber is electron.
1720s C.E.	Static electricity machines are built for fun and experimentation.
1790s	Claude Chappe develops a semaphore signaling system.
1800	Humphry Davy sees that electricity is a chemical reaction.
1824	Sadi Carnot puts forward a theory of thermodynamics.
1830s	Samuel F. B. Morse develops the telegraph.
1831	Michael Faraday invents the dynamo.
1840s	The laws of thermodynamics are worked out.
1860s	James Clerk Maxwell postulates the electromagnetic wave theory.
1876	Alexander Graham Bell invents the telephone.
1891	Telefon Hirmondó begins service over telephone lines in Hungary.
1893	Nikola Tesla lectures on wireless communication.
1894	Heinrich Hertz shows that electromagnetic waves can be generated and shown with a spark gap machine.

1896 Guglielmo Marconi gives his first demonstrations of wireless signaling. Jagadis Chandra Bose does similar experiments.

1901 Marconi sends successful transatlantic wireless signals.

1904 Hugo Gernsback is selling Telimco wireless sets commercially. John Ambrose Fleming invents the vacuum tube.

1905 Lee De Forest improves the vacuum tube and calls it an audion.

1906 Reginald Fessenden makes the first audio wireless broadcasts. The crystal radio set is born.

1909 The Marconi Institute opens in New York City to train wireless telegraphy operators. (The school is changed in the 1920s to the RCA institute, then in 1971 to the TCI College of Technology.)

1912 The RMS *Titanic* sinks. David Sarnoff plays a role in transmitting names of the lost. The Radio Listening Act is passed in the United States.

1914 American Society of Composers, Authors, and Publishers (ASCAP) is formed. The Canadian government forces amateur wireless operators off the air.

1915 Lee De Forest's Highbridge radio station begins broadcasting.

1916 David Sarnoff writes the Radio Music Box memo while with American Marconi.

1917 The Bolsheviks of Russia begin political broadcasting. The U.S. government forces amateur wireless operators off the air.

1918 Edwin Armstrong invents the heterodyne circuit.

1919 American Marconi is reformed into Radio Corporation of America (RCA).

1920 Frank Conrad's KDKA begins broadcasting out of Pittsburgh. Marconi sets up station 2MT in England.

1921 Amateur radio begins in India.

1922 Wendall Hall becomes the first paid regular performer on radio.

1923 The *EverReady Hour* and the *Roxy and His Gang* show begin to air. The National Association of Broadcasters (NAB) is formed. The Operadio portable radio is available.

1924 Aimee Semple McPherson begins broadcasting. WGN (the World's Greatest Network) goes on the air in Chicago.

1925 The Chicago Cubs baseball team begins airing live play-by-play. The Scopes's "Monkey Trial" airs over WGN.

1926 Father Coughlin begins broadcasting. The British Broadcasting Corporation (BBC) is formed.

1927 The Federal Radio Commission (FRC) is formed in the United States. The NBC and CBS networks are formed. William Paley takes over CBS.

1928 *Amos 'n' Andy* debuts.

1929 Radio Normandie begins broadcasting in France.

1930s FM radio is created by Edwin Armstrong.

1933 Franklin Roosevelt gives the first Fireside Chat. Adolf Hitler makes his first radio broadcast.

1935 Station VU2HK begins in Mysore, India.

1937 The FRC becomes the Federal Communications Commission (FCC).

1938 Orson Welles puts on the *War of the Worlds*.

1939 Broadcast Music Incorporated (BMI) is formed.

1940 Edward R. Murrow does his *This is London* broadcasts. The ABC network is formed. The first FM station is set up in New Jersey.

1941 *The King Biscuit Time* show debuts.

1945 Barry Gray holds the first on-air conversations with listeners; talk radio is born.

1947 The transistor is invented.

1949 Alan Freed begins his rock 'n' roll career. *Pacifica Radio* goes on the air in California.

1950s Rock 'n' roll begins to dominate radio airplay.

1957 Sony releases the TR-63 transistor radio.

1957 Screamin' Jay Hawkins's "I Put a Spell on You!" is banned from radio play.

1959 The integrated chip is invented.

1960s The Payola scandal hits the U.S. music and radio industry. Top 40 is invented in response. FM returns as a viable broadcast medium. The first all-talk stations appear. Roth-Memoirs test for obscenity is adopted by the U.S. government.

1963 "Louie, Louie" is investigated by the Federal Bureau of Investigation (FBI) for obscenity.

1964	*Jacobellis v. Ohio* case on obscenity in the United States. Potter Stewart says of obscenity, "I know it when I see it." Marshall McLuhan publishes *Understanding Media.*
1978	*FCC v. Pacifica* case deals with the words you cannot say on radio or television.
1980s	Radio deregulation allows for the growth of large radio networks. Talk show hosts Bob Grant, Rush Limbaugh, and "shock jock" Howard Stern gain their notoriety.
1985	The Fairness Doctrine is gutted.
1986	The Canadian government passes laws prohibiting obscenity.
1990	2 Live Crew is deemed obscene by a Florida judge.
1990s	The FCC fines Howard Stern for vulgarity. The French government restricts how much non-French content can be on its airwaves.
1996	The Telecommunications Act is passed to deal with child pornography and indecency. Radio consolidation begins.
Early Twenty-first Century	Satellite radio begins. Internet simulcasting.
2004	*Air America Radio* goes on the air.

1

The Ancestry of Radio

Radio, like any technology, did not just appear overnight as a complete system. Two centuries of discoveries, theories, and scientific principles in seemingly disparate areas preceded it. Almost none of the researchers discussed in this chapter were thinking of radio as they did their work. Some were interested in putting together devices that could make them rich, while others were interested in nothing more than working out abstract ideas like the nature of lightning or the properties of electromagnets. Later inventors and theorists built on what these predecessors did. While many areas of research contributed to the development of radio, the most logical starting point for the discussion is the history of electricity. Eighteenth-century natural historians and philosophers (as scientists were then called) were fascinated by the elusive agent they came to call—electricity. Their research, which began as an attempt to understand the nature of this quirky phenomenon, led to a deeper understanding of the very workings of the universe as well as a way of sending sound over the airways. The ability to create, use, and control electricity would be crucial to the development of radio.

ELECTRICITY

The earliest knowledge of electricity may be from 600 B.C.E. when the Greek philosopher Thales of Miletus saw that amber would attract small objects to it after it had been rubbed vigorously against cat fur (the Greek word for amber is electron). What is considered the first milestone in the study of electric effects came in 1600 with the publication of British court doctor William Gilbert's *De Magnete*. Gilbert (1544–1603) was a physician to the Queen of England who did experiments with electricity and magnetism. He showed that besides amber, materials such as glass, diamonds, and the like became magnetic when rubbed. He dubbed the attractive power of magnets electrica (the term *electricity* did not come into use until the 1640s). The word for magnet is thought to trace back to a Greek legend of a region known as Magnesia where, it was said, certain stones could pull the iron nails from the sandals of anyone walking over them. Gilbert noticed that naturally occurring magnets known as lodestones did not need to be rubbed to attract objects, and it only attracted metallic ones. At the same time, rubbed or *electrified* bodies would attract anything that was light enough. Also, he saw that magnetic bodies attracted objects in a certain pattern while electrified bodies just piled them up. He felt that the electric properties of the object were inherent in the object and were released by rubbing.

The scientific and popular appeal of electricity in the West picked up in the 1720s when more practical devices for generating electricity were developed. These were typically cranked affairs that generated static, which could be siphoned off for other applications. Electricity was not just an intellectually interesting phenomenon, but an amazing spectacle. Giving public demonstrations of electrical effects became a rage in Europe. When they were not investigating the workings of electricity in their laboratories and workshops, some researchers hired themselves out for the amusement of paying guests. They would perform experiments that showed the unusual behavior of electricity as well as other phenomena then only crudely understood. Partygoers would squeal in delight as sparks were made to crack and jump, and even give shocks as entertainment.

The person who embodied the serious study of electricity during the eighteenth century was the American autodidact Benjamin Franklin (1706–1790). Though not a trained scientist, Franklin had a voracious interest in the natural world. Franklin's initial interest in electricity was, as it was for many in the early days of electrical research, from an entertainment point of view. However, as he investigated more, he became increasingly serious. His first exposure to electricity came in 1743 when he witnessed an electric show put on by Scottish demonstrator Dr. Archibald Spencer. So taken was Franklin that he immediately began experiments of his own and even acted as Spencer's agent briefly. Franklin drew some of his intellectual Philadelphia friends into experimenting with him and had local artisans

Pieter van Musschenbrock, Dutch physicist, demonstrates the principal of the Leyden jar, April 20, 1746. The Leyden jar became a fixture of electrical experimentation for over a century. Undated Woodcut. Copyright Bettman/CORBIS.

construct instruments to aid in their investigations. As a result, he determined that electricity was not being collected, but formed by the experiments and that lightning was a form of electricity. In his work Franklin employed an early electrical device known as the Leyden jar. This was a glass jar wrapped in foil and wired in such a way as to be able to store static electricity introduced into it. The Leyden jar became a popular experimental device and remained a fixture with experimenters into the nineteenth century. Franklin saw that the outside and inside of the jar were oppositely charged. He built a more sophisticated Leyden jar utilizing a series of glass plates wired together. As his contraption was arranged with the plates in a row like guns on a ship, he called it a battery.

SIGNALING

The need to be able to communicate information over long distances was appreciated early on in the history of civilization. The first attempts at such long-range communication involved persons physically carrying messages from one spot to another. The Greek story of Marathon is only the most well-known example of such human-powered communications. Societies and governments worldwide employed some type of runner system to bring news from one end of an empire to another. Early signaling devices usually employed flags or smoke. Signalers would be stationed on mountains or towers within sight of each other. The Romans did this as did the Chinese on their Great Wall. These techniques, while crude, could transmit messages effectively enough to allow these empires to prosper and spread. Speedy communication was essential to a government's survival and new and better methods were always looked for.

In the 1790s, Frenchman Claude Chappe (1763–1805) became interested in developing a messaging system that utilized electricity, but was unable to work out an idea. Instead, he tried a manual system. Chappe was a priest who lost his position after the French Revolution and turned to engineering. He and his brothers tried a series of complex designs involving sound created by specially made clocks, bells, or clanging gongs. As a result of its limitations, they abandoned sound and switched over to a visual system and hit upon the idea of building a tower with movable arms. The positioning of the arms corresponded to parts of a special code. In this way a complex message could be sent with simple signals. The visual approach was far more successful, so Chappe planned on expanding the system. He originally called his semaphore device a *tachygraphe* (fast writer), but then changed it to *télégraphe* (far writer). A "telegraph line" was set up between the cities of Lille and Paris and could send a message in one hour that would have taken dispatch riders 24 hours to convey. After seizing power in 1799, Napoleon Bonaparte ordered the expansion of the system, eventually covering 3,000 miles with over 550 individual stations. Napoleon kept the system in the hands of the government with private messaging systems outlawed. As a result, few people had direct access to such forms of communication. While this improvement in signaling spread to other countries quickly, Chappe's semaphore had limitations of its own: it could not be used at night or in bad weather. Other nations, the British for example, developed similar devices. During the Industrial Revolution a railroad signal system, very much like Chappe's semaphores, was adopted. The original semaphore was still being used as late as the Crimean War in the 1850s because on a battlefield, it could be set up and used faster than a wire telegraph. Though he did receive recognition for his work and a level of success under the French State, Claude Chappe grew despondent and, after a bout of depression, committed suicide in 1805. Despite the success of the visual semaphore system for communicating information across distances, the

The French visual semaphore was the first high-speed communications device. Note the flexible arms on top of the tower. Copyright Stefano Bianchetti/CORBIS.

pursuit of the idea of an electric telegraph intensified. As knowledge of electricity grew, some thought it might be possible to send not just current across a wire, but intelligible signals. Throughout the early decades of the nineteenth century, a number of experimenters developed electrical signaling devices with varying levels of success.

THEORETICAL BREAKTHROUGHS

Along with the practical work, a number of discoveries were made about the underlying nature of electricity, which were needed for the development of radio. By 1800, it had been discovered that certain materials broke down or decomposed when exposed to electrical current. Humphry Davy (1778–1829) advanced this knowledge when he built an enormous 889-square-foot battery to explore the phenomenon. He saw that this process, later called electrolysis, was a chemical reaction and therefore the constituent parts of chemicals were held together by electricity. Davy became a respected speaker at the prestigious Royal Institution in London. During his last lectures in 1810, a young man named Michael Faraday sat in the audience. Faraday (1791–1867) was a largely self-taught scientist who had become fascinated by electricity and transfixed by Davy's lectures. Starting off as a bookbinder's apprentice, Faraday learned a great deal about science by reading the books he was binding. In an effort to impress Davy, he bound up the copious notes he took during Davy's lectures and sent them to him as a present. He also asked Davy for a job. Davy was initially uninterested, but acquiesced after he suffered a laboratory accident and needed to take on an assistant. Moving out from under Davy's shadow, Faraday (who eventually took over Davy's old job at the Royal Institution) made a number of important discoveries. Possibly the most important was that of induction: the creation of electricity in a wire by using magnets. In 1820, Danish physicist Hans Christian Oersted (1777–1851) showed that passing a current through a wire created a zone of effect around it called a magnetic field. The presence of this field could be detected by its effects on objects around it, like a compass needle. The electrified wire not only created a field of influence around it, but also acted like a magnet. A moving needle became the heart of a device designed to detect magnetic fields, the galvanometer. While Oersted showed that electricity could create a magnetic field, many argued that logically the reverse should also be. The problem was that no one was able to prove it. Faraday worked out the problem in 1821 when he connected a galvanometer to an iron ring wound with wire and connected to a battery. What Faraday did differently from previous experimenters was that he attached the detector to the coil *before* attaching the battery. This allowed the detector to register current. Continuing his work, by 1831 Faraday had built the first dynamo, or electric generator.

By the 1840s, Faraday came to believe that electricity, light, and heat were an effect of electromagnetism and wanted to put together a single theory that would account for it. He also thought that this unified relationship was an extension of God's unity of creation. James Clerk Maxwell (1831–1879) built upon Faraday's idea and became one of the great theorizers of the nineteenth century. He worked on a range of topics including the nature of color vision and photography: producing the first color photograph. He also determined that all colors were a combination of red, green, and blue. He did his work on electromagnetic field theory in the 1860s, showing that energy came in waves and that Faraday had been correct in thinking that different forms of energy were really just manifestations of the same thing. He also argued that a wire was unnecessary for these waves to travel from one place to another, and posited that light was only one part of an entire spectrum of electromagnetic waves.

The nature of heat and energy were also being worked out during this period. In the 1820s, a French military engineer, Sadi Carnot (1796–1832) was working on ways to improve steam engines by understanding the theoretical foundations for the movement of heat. His *Reflections on the Motive Power of Fire* (1824) laid the framework for what would be known as thermodynamics. He said that under normal conditions heat always travels from a warm to a cool body. Unfortunately, Carnot's book was not widely read at first. Eventually, others began to make discoveries about how heat and energy behaved. In the 1840s, Julius Mayer, James Joule, Hermann von Helmholtz, and others worked out the idea that energy could neither be created nor destroyed, but could only be converted into other forms of energy. Also known as the conservation of energy, these ideas were codified as the laws of thermodynamics. In the next decade, German physicist Rudolf Clausius (1822–1888) advanced Carnot's work by arguing that heat does not travel spontaneously from a cooler to a warmer body and that heat/energy has a tendency to weaken or dissipate, and thus coined the term "entropy" to describe this phenomenon. These theoretical ideas would become crucial for the advancement of mechanical science and be necessary for the invention of the telegraph and radio.

The work on electromagnetism done by Maxwell was advanced by the German physicist Heinrich Hertz (1857–1894). Of all the theoretical work discussed in this chapter, Hertz's is the one which can be traced most directly to the invention of radio. In the late 1880s, Hertz was trying to prove in reality what Maxwell had suggested in theory. In his laboratory at the Karlsruhe Polytechnic, Hertz built a machine to test for the presence of electromagnetic waves. The device comprised of two metal rods lined up end to end but with a space, or gap, between them. Current was then introduced into the rods (each receiving an opposite charge) causing a spark to jump the gap. A similarly constructed device could then register the electromagnetic wave as it was being generated. The oscillating nature of the wave was evident and thus proved its existence. With this device Hertz had shown that waves could be artificially produced and measured, and suggested that light and electromagnetic waves traveled at the same speed and that electricity could be transmitted without the need of a wire. Hertz published his findings in an article in the German physics journal *Annalen der Physik* shortly after. It was this article that was read by a young Italian named Guglielmo Marconi and started him thinking about a practical application for the idea.

THE TELEGRAPH AND THE TELEPHONE

While theory was being debated and pondered, practical experimentation continued. Sending an electric current across a short length of wire was one thing; sending that current over many miles was another. After a few hundred feet, the current just seemed to dissipate and vanish. Throughout the early nineteenth century, a number of inventors and scientists sought a way to make an electrical telegraph. Most failed because of the obstacle of range. In the end, a number of people came up with bits and pieces of hardware and solved various problems separately, which finally came together in one system. The Englishman William Fothergill Cooke (1806–1879) built an improved version of a telegraph design of Russian inventor Baron Pavel Lvovitch Schilling (1780–1836). The Russian devised a working model of a telegraph and received backing from the Czar to develop it into a full-scale program. Unfortunately, Schilling died before the project started and it was dropped. After leaving the Indian Army in 1831, William Fothergill Cooke returned to university studies in Paris where he saw a demonstration of the Schilling telegraph and became fascinated by it.

Unbeknownst to amateurs like Cooke, the problem of how far current could travel along a wire had been addressed by American physicist Joseph Henry (1797–1878). Henry was a prominent scientist who did pioneering work on electromagnets, became secretary of the newly formed Smithsonian Institution in Washington, D.C., in 1846, and was a major supporter of science studies in America. Englishman William Sturgeon had invented the electromagnet: a horseshoe-shaped iron bar wrapped with wire. The iron magnetized when a current was passed through it and demagnetized when the current was switched off. Henry improved on Sturgeon's electromagnet by wrapping the wires around the iron bar in tighter coils. He also saw that there was an optimum amount of wire which produced the greatest power. In his investigations into the problem with range, Henry saw that the answer to sending electricity along a wire was not in the wire, but in the battery. He solved the problem by using a series of smaller batteries instead of a single large one. He published his work in the *American Journal of Science* in February of 1831.

In England, William Fothergill Cooke managed to get an interview with Michael Faraday to discuss his work. After initial interest, Faraday distanced himself from Cooke thinking him too much of an eccentric: in addition to the telegraph, Cooke tried to sell Faraday on a perpetual motion machine. Cooke was eventually introduced to the scientist Charles Wheatstone who had already managed to send a current over 4 miles of wire using the principles of Joseph Henry. Cooke and Wheatstone saw how each could benefit the other and joined forces. They proposed an electrical signaling system to several British railroads. The enthusiasm shown by the companies allowed the two men to continue their work. While Cooke and Wheatstone never meshed personally, they did advance telegraph technology together. Wheatstone, as a trained scientist, felt that he should get credit for inventing the telegraph and saw Cooke as merely a financial backer. The two argued over invention credit the rest of their lives. What they made was a further improvement on the Schilling system using needles pointing to letters and numbers. However advanced their telegraph was, it was still an awkward system where not all the consonants of the Latin alphabet could be represented.

At about the same time in America, the artist Samuel F. B. Morse was working on an idea of his own. Morse (1791–1872) studied art in Europe and became a skilled portrait painter. He was a staunch nationalist and harbored desire for elected office. His interest in the telegraph began in 1832 during a sea voyage home from Europe. During the trip, he fell into daily conversations with fellow passenger Dr. Charles Jackson. Hailing from Boston, Jackson was fascinated by electromagnetism and was even carrying several instruments with him onboard. Intrigued by Jackson's knowledge of the subject, Morse was struck by the idea of building a telegraph. He was unaware of the work of Henry, Cooke, and others, and so thought his idea new and untried. In addition to being a painter, Morse was an amateur inventor and schemer. He came up with a number of novel ideas for mechanical pumps as well as a device for making exact copies of statues so that inexpensive versions of famous marbles could be sold to art lovers. During that fateful voyage, Morse threw himself into a fevered attempt to design a telegraph. Most of his time, however, was devoted to working out an efficient code system. He worked through a number of ideas until he settled on a series of dots and dashes made by alternately sending current through the line. He also devised a printing system to record messages as they came and went. Upon his arrival home, Morse corralled his brothers into his new venture, but suffering from limited funds, their work progressed slowly over the next few years.

Morse's artistic career prospered and New York University (NYU) offered him a permanent teaching position. While the pay was modest, it gave Morse a studio where he could continue working on the telegraph. At NYU a chemist named Leonard Gale became interested in Morse's work and suggested that they switch over to using Joseph Henry's design for electromagnets for power. This improvement allowed them to send messages over 10 miles of wire. A short time later, Morse and Gale were joined by Alfred Vail, a young man who had seen Morse give a demonstration and wanted to get involved. Vail's money allowed Morse to construct a full-scale working model of his telegraph, for which he was given a share in the patent rights. With this new impetuous, Morse's work sped up. They dropped the cumbersome wire and letter notice board for a simpler key pad, which would tap out the code. Morse also started corresponding with Joseph Henry, who helped Morse acquire federal funding for the telegraph project in 1842. Henry was a respected scientist and felt Morse's design the most promising of all the models then being worked on. However, Morse, Henry, and others were later caught up in a battle over the patent rights to this lucrative invention. While Henry never claimed to have invented the telegraph—only crucial parts needed for its success—he grew annoyed with Morse. Henry, the professional scientist, felt Morse, the amateur, was hogging a bit too much of the glory. Henry just wanted acknowledgment for his contribution and Morse was reluctant to give it. As with many inventions, radio included, a long and bitter struggle ensued over "who invented it?" While many worked on the problem and made important strides toward its completion, what is clear is that, in the end, the telegraph most widely used was the Morse model.

Throughout his life Morse showed a dark side with his involvement in the pro-slavery and anti-Catholic movements. Morse was convinced, as were many in nineteenth-century America, that a great plot to overthrow the Republic was being hatched by the Pope and Catholics living in the country. He joined growing ranks of those who railed against this plot and called for its elimination. He saw priests and nuns as especially dangerous to America's future. Writing under the pen name "Brutus," Morse published *The Foreign Conspiracy Against the United States* (1835). In the introduction, he suggested that anti-Catholic hatred was actually a charitable thing. He also supported the Southern cause in the Civil War and attempted to "prove" that slavery was meant to be. In *An Argument on the Ethical Position of Slavery* (1863), he claimed that the duty of man was obedience to God. In logic, that would have made Thomas Hobbes proud, as he argued that there were certain relationships that man must follow. These institutions included the ruler and the ruled, husband and wife, parent and child, and master and slave. Therefore, Morse argued that it was foolish to oppose slavery because it was biblically mandated and a perfectly normal course for a Christian to follow. Morse also felt that it was the forces of abolition and not slavery that were responsible for the war. He saw abolitionists as fanatics who were going against the will of God, and did not have a high opinion for anyone engaged in foolish "humanitarian philanthropy."

Samuel Morse knew there was going to be something needed more than just a machine and a code. It did not matter how well a telegraph tapped out messages if there was no one to hear them. What was required to make the telegraph a financial success was a network. Morse and the others also needed to convince people of the practical value of such a network. After finishing the rounds of Washington, D.C., he went to England where he ran into Cooke and Wheatstone with mixed results. The two Britons had worked out a deal with the London and Birmingham Railway to allow them to put up a telegraph line between Euston and Camden Town stations. Morse then hit the continent with even less promising results. He and his financial backers began to despair when the U.S. Congress came through. With this backing, Morse was able to put up a line between Washington and Baltimore, and he used it to announce the nominees of the 1844 Whig Party political convention. Shortly after, in England, the telegraph was used to announce the birth of Queen Victoria's child, Alfred Ernest. Events like these helped spread interest in the telegraph as a viable system of information transmission. Morse found greatest support in the private sector, sending individual messages for a fee. This led to the creation of a special telegraph system dedicated to sending business information on the sales of stocks and bonds. The telegraph grew so quickly once its usefulness was recognized that by 1852 *Scientific American* could say that no other invention had grown so quickly (in 1850 there were over 12,000 miles of lines). By the 1860s, they stretched from the Atlantic seaboard to the West Coast. It grew so rapidly that by the time of the American Civil War, the telegraph was increasingly seen as an integral part of American business.

The signal telegraph train as used at the battle of Fredericksburg during the American Civil War. Communications were vital to the prosecution of the war by both rebels and government. Courtesy of the Library of Congress.

By the mid-nineteenth century, telegraph lines were spreading over not just the United States and England but Europe, South America, and Australia.

The telegraph allowed for the rapid dissemination of information and began the first worldwide communication network. It opened up economies and helped spread national influence around the world. It also made the world a smaller place as buzzing wires and tapping receivers drew small communities and entire countries closer together. In the United States, the Union and Confederate governments relied heavily on the telegraph to prosecute the Civil War. The spread of the British Empire was assisted by it, and even after the advent of radio, remote parts of the world were still using it into the mid–twentieth century. It opened communications so much that historian Tom Standage called it the Victorian Internet. As important as the telegraph was, some dreamed of giving birth to better systems. The telegraph had certain drawbacks; knowledge of Morse code was needed to understand the messages. What would be easier would be a device that allowed people to communicate without all that technical knowledge: to communicate by talking instead of tapping.

Alexander Graham Bell (1847–1922) was born in Scotland into a family that was interested in interpersonal communications. His grandfather was an actor who also taught speech and published a book in the 1830s to help stammerers overcome their difficulty. Bell's mother, Eliza Grace Bell, was deaf and their relationship led young Alexander to believe that the deaf might be able to hear through vibrations in the body. After tuberculosis claimed the lives of several family members, Bell's parents took him to Canada. By 1871, Bell was in Boston teaching speech at a school for the deaf. He was convinced that the deaf should be taught to speak so that they could survive more easily in the mainstream world of the hearing. He was opposed by those who felt just as strongly that the deaf should be taught sign language instead. Bell married one of his students, Mabel Hubbard, and continued to think about how to help the deaf hear.

As a student at the University of London, Bell had become acquainted with the writings of German physicist Hermann von Helmholtz, who was involved in the work on thermodynamics and Maxwell's electromagnetic theory. In 1863, von Helmholtz published *On the Sensations of Tone as a Physiological Basis for the Theory of Music,* in which he said that musical tuning forks could reproduce certain qualities of human speech. Bell's command of the German language was not thorough and, as a result, he misinterpreted a part of Helmholtz's work as saying that human speech could be transmitted over a wire. During his years in Boston, Bell began working on what he called a harmonic telegraph that could be used by the deaf to enhance their ability to communicate. He met and began to collaborate with a local scientific instrument maker, Thomas Watson (1854–1934). In 1871, Watson was adjusting a device they were working on when his tinkering produced a sound, which came out of the other end of it. He had inadvertently done what they were trying to do: send a sound over a wire. In 1876, Bell patented the device even though he had yet to make it work transmitting a human voice. Bell and Watson had seen that a human voice could make a wire vibrate if it were suspended in a fluid that acted as a conductor. Just a few days after having received the patent, they unexpectedly met with success. According to legend, Bell and Watson were conducting experiments with their machine wired in two different rooms. The fluid they were using for the conductor was highly corrosive battery acid. Bell accidentally spilt some on his hand, causing an immediate and painful burn. Bell gasped out something like, "Watson, I need you!" In the next room, Watson heard the plea over the device. They had made the telephone work.

Bell immediately began promoting the telephone, also putting it into the Philadelphia Exhibition later that year and wiring a telephone into the White House in 1878. This allowed Rutherford B. Hayes to become the first American president to make a phone call. The Western Union Company, which had taken control of a large chunk of the telegraph industry, wanted to control the telephone as well. To that end, they engaged Elisha Gray of Chicago and Thomas Edison to work on the project. By 1877, Edison improved Bell's design by using a carbon disk in the transmitter to give a much clearer sound. Bell's company, Bell Telephone, sued Western Union over patent infringement and won (Bell Telephone eventually became AT&T). Thomas Watson had no interest in running a company or even furthering his work on the telephone. By his late twenties, he embarked on a colorful career (which included shipping yard owner, mining engineer, and finally touring stage actor), backed up by telephone royalties.

THOMAS EDISON AND ELECTRIFICATION

It is impossible to discuss the electrification of the modern world without mentioning Thomas Edison. Neither a trained scientist nor an engineer, Thomas Alva Edison (1847–1931) was a classic inventor. He placed emphasis on nuts and bolts, not the temporal gymnastics of theory. His original factory in New Jersey was directed at designing working prototypes, not for manufacturing finished products. His equation about success placed 99 percent on perspiration (dogged experimentation) and only 1 percent on inspiration (theory). He cared less why something worked than that it did. His work went toward producing practical results, and he had an innate understanding that to sell technology to the public would require gadgets with entertainment as well as practical value.

As a teenager living outside Detroit, Michigan, Edison learned to be a telegraph operator. During the Civil War, experienced telegraphers were being siphoned off for the war effort by both sides. This left the field open for Edison to begin building a career. Edison read voraciously and did his own experiments. As was common for telegraph operators, he traveled widely as a kind of itinerant technician. By the late 1860s and early 1870s, Edison was inventing machines, including an improved telegraph and stock ticker. He also entered into different business ventures and established workshops to sell and manufacture equipment with mixed results. In 1875, he opened a factory at Menlo Park, New Jersey. It was the first industrial research facility where he brought together a number of specialists to tackle different design problems. Prior to Edison, invention tended to be a haphazard affair with results coming down in many cases to luck. Edison's idea was to concentrate energy on specific problems and keep at them until an answer was found. In this way, invention ceased being a hobby and became a method and a profession. His attitude of single-minded perseverance is possibly best illustrated by his work on the light bulb. While others had been working on artificial lighting, Edison solved the problem by creating a vacuum inside the bulb. Humphry Davy had shown that certain materials would glow, or incandesce, when electricity was passed through them. Edison tried hundreds of different materials for the filament before adopting one made of carbon.

Edison's success with the incandescent bulb gave a viable artificial light source. The problem was that there was no electrical system to attach it to. Besides being an inventor, Edison was also a thoughtful and scrupulous businessman. He knew that all the electric devices in the world meant nothing without an electrical system and so set about electrifying the world. To this end, Edison's research team led by Francis Upton produced an improved dynamo (many of the designs credited to Edison for "inventing" were actually the products of members of his team). A substantially sized generator would be needed for commercial purposes. The men quickly nicknamed the new big generator "Long Legged Mary Ann," which was not only the largest generator built up to that time, but also the most efficient. In order to spread the gospel of electrification, Edison engaged in a number of public relations stunts from lighting his factory to lighting up a cruise ship and sending it on a trip around the Americas. He also built an electric railroad on the grounds of his factory. He set his sights on the electrification of New York City as the ultimate advertisement for his services. He was able to persuade city officials to let him set up a small pilot program in a ten-block area centered around Pearl Street on Manhattan's Lower East Side. A number of competing companies had set up similar operations in New York, all hoping to corner what was to be a very lucrative financial market. Inventors Hiram Maxim (inventor of the machine-gun) and Joseph Swan had put up incandescent lighting systems in England, but Edison's overshadowed theirs.

Edison's greatest rival was George Westinghouse (1846–1914), who gained fame in the 1860s for inventing air brakes for trains. Westinghouse wanted to get in on the fledgling electrical industry and planned on using Alternating Current (AC) for the job. Edison wanted to use Direct Current (DC). AC differed from DC in that it could flow in both directions over a wire. This allowed for much higher voltage and, more importantly, allowed current to be transmitted over greater distances. DC could only be sent a half mile or so. This meant that a DC system would require many generating stations located alongside populations and businesses. AC would allow the generating facilities to be located far outside urban areas and need fewer stations to supply power to a given geographic region. To turn potential customers off AC, Edison had a number of grisly experiments performed where animals were electrocuted to show the lethal nature of mishandled AC power. He also designed a device to execute criminals (the "electric chair" was one of Edison's less savory inventions). Despite Edison's opposition to it, AC became the standard system allowing electrification to spread quickly throughout America and then the world. For example, in 1882 there was only one electric generating plant in the United States; by 1902 there were over 2,000. Thomas Edison's New York operation became the Consolidated Edison Company (ConEd) and the bulk of his research and manufacturing facilities were brought together by financier J. P. Morgan as General Electric (GE). Following the formation of the Bell Telephone Company in 1876, long-distance calling was available by the 1890s, linking cities hundreds of miles apart. The spread of this system led to the growth of the modern corporation, which in turn led to developments of new technologies in an upward spiral of design, integration, and growth. By the end of the nineteenth century, 70 percent of American homes were wired for electricity. Morse's telegraphs, Edison's lights, and Bell's telephones had become inseparable parts of American and European and, increasingly, world business and society.

CONCLUSION

In the rapidly industrializing Western world people were growing less dependent on nature and more dependent on other people for survival. They were becoming more and more attached to technological systems: complex interplays and interdependencies of linked technologies. For example, to perform their jobs at night, office workers relied on electric lights that were kept running by generator workers and operators, who relied on field technicians to keep the lines up and functioning. Electric elevators were needed to bring people to upper floors and electric trains to take them home. One individual's life and career required hundreds of other workers to keep it going. A break in any part of that system could adversely affect the office worker's life and there was nothing the office worker could do about it. Part of this systemization of technology was that, as electrical technology became more prevalent and important in people's lives in the late nineteenth and early twentieth centuries, it became increasingly invisible. Users of technology were less interested in the technology for its own sake and more interested in what it allowed them to do. The pervasiveness and ubiquity of electrical technology bred a familiarity, which removed much of the novelty from it. The awe-inspiring "wonder" of the early eighteenth century had been replaced by a mundane usage of the twentieth century. The same thing would happen when radio passed from being a machine to being a media.

The telegraph was the first form of long-range electric communication.

In some ways, the history of radio mimics that of the telegraph. Both had rocky starts as technical problems were worked out, both were employed initially by the military, and only later came into public use. Beyond that, however, radio was different. The telegraph and telephone were personal communication devices: one person sending a message to another. Radio was a more passive agent of communication. Except for individual enthusiasts, radio messages were sent to an anonymous mass audience, not individuals. While the telegraph laid down important technological groundwork for radio, it was the newspaper that was the philosophical ancestor to radio. The telegraph was a communication source to the masses only when its messages were sent to newspaper offices, which then printed them for distribution. Radio advanced the communication revolution begun by printing in a radical way. All that was still in the future, however. At the close of the nineteenth century, radio was still a dream. That was about to change. The parts needed for radio were mostly in place: batteries, electric generators, wave theory, and just as importantly, a population growing used to technological marvels and keen to use them in their lives. The nineteenth century was about to give birth to the twentieth, and the first birth cry would be over the radio.

2

Radio Is Born

By the 1890s, most of the elements necessary to give birth to radio were present: electricity was fairly well understood; the telegraph was widespread and the telephone was becoming so; thermodynamics and electromagnetic theory were being studied both theoretically and practically. While no concerted effort was being made in that specific direction, a number of people were working on different aspects of what would become *radio*. The concept of radio, as we understand it today, had not yet coalesced into a reality. What was on most people's minds was wireless telegraphy. A question often asked today is, "Who invented radio?" This is a misleading question as no one person did. Radio was both a conscious and unconscious group effort. While the Italian Guglielmo Marconi would receive credit as the "Father of Radio," there were many people who contributed to its conception.

NIKOLA TESLA

The history of radio, or just about any modern electrical technology for that matter, must discuss the contribution of Nikola Tesla (1856–1943). One of the more enigmatic and tragic figures of late-nineteenth and early-twentieth-century science, Tesla possessed a prodigious intellect that could visualize an astounding array of ideas and concepts in his mind that are still being extrapolated decades after his death. Thomas Edison thought him a poet. His ideas were so revolutionary and futuristic that an early biographer was convinced he was an extraterrestrial from Venus. A more prosaic account called him a "prodigal genius." Though ethnically Serbian, Tesla was born in the Croatian village of Smiljan. Tesla's father was a stern minister and his mother, an intelligent but uneducated peasant woman who taught her son to memorize things and read as much as possible. While still a child, he began to invent devices including a paddleless waterwheel. His eccentric personality may have had its roots in the clouded, accidental death of his elder brother Daniel whom Nikola admired and over whose death felt some guilt. His mental condition was such that throughout his life Nikola experienced flashes of bright lights before his eyes. Despite this, he developed a powerful ability to visualize designs, schematics, and technical drawings in his mind without the need to write them down. Throughout his career, he amazed and frustrated his colleagues by having complete plans for inventions in his head but not down on paper. While he built many working prototypes and had a mechanic's ability to fix broken machinery, he was a consummate theoretician who focused on novel explanations and applications for technology.

In the mid-1870s, Tesla began a rigorous education at the Austrian Polytechnic School in the city of Graz. It was here that he first became interested in working out the problems of alternating current (AC). He eventually produced a motor that used a rotating magnetic field produced by two electric currents running out of synch with each other (other engineers had used only one current). Several inventors had begun to produce AC motors around that time, but they were unreliable and prone to breakdown. By the late 1870s, a number of events conspired to eliminate Tesla's scholarships and so he was forced to leave school. While he continued his education with intensive self-study, he had to find work to support himself. He soon found it at Thomas Edison's European subsidiary in Paris as an engineer. He also tried unsuccessfully to find financial backers for his AC project; the Edison Company, like its namesake, was wedded to direct current (DC). Having impressed his superiors at Edison, however, he was prompted to go to America and work for the great man himself. He duly sold all his possessions and left for the New World, landing in New York in 1884 with a few dollars in his pocket, copies of some of his articles, and a head full of dreams.

Tesla showed up at Edison's office for his interview, with the office in the middle of a crisis. Tesla jumped right in and helped solve the situation. As a result, Edison gave Tesla a job working on dynamos. The two, while respecting each others talents, did not get along. Edison was the dogged, crude, self-taught inventor, Tesla the sophisticated intellectual. Tesla saw Edison's legendary technique of working through every problem one step at a time a waste. He argued that theoretical analysis would have eliminated a good portion of wasted effort. They were soon at odds with each other. Tesla felt Edison lied to him and cheated him on a business deal, and so he quit. He then went to work for George Westinghouse, who was locked in the battle over who would get the lucrative contracts to electrify New York City. Westinghouse had been trying to make AC work without much luck. He immediately saw the efficacy of Tesla's designs and brought him on. Eventually, Westinghouse purchased Tesla's AC patents and gave him a hefty salary.

In the early 1890s, Tesla began promoting the idea of electric motors that could run without wires. He was convinced that current and energy could be transmitted through space. He even suggested that power need not be transmitted at all. Electrical devices could draw on the energy in the surrounding atmosphere. He eventually developed the "Tesla Coil," which was a series of coils tuned to each other and a "step-up transformer" to convert low-voltage into high-voltage. It was the beginning of the type of high-voltage alternator needed for continuous-wave radio communication. During this time Tesla's work was becoming widely known. In 1892, the British physicist Sir William Crookes (1832–1919) gave a lecture based on Tesla's work, in which he said that electromagnetic waves could be used to make a wireless telephone or telegraph. In a pair of 1893 lectures, Tesla himself laid out what he considered the principles of wireless communication. In St. Louis in 1895, he demonstrated these principles for the first time with a transmitter and receiver set up on stage. The device contained a power source, an aerial wire, a ground, and the ability for each device to be tuned to the same frequency. Tesla, however, did not transmit a voice, but vibrations.

Around the same time, Sir Oliver Lodge (1851–1940) showed that telegraph signals could be sent by wireless transmission up to 150-yards distance. Lodge was a physicist at the University College of Liverpool, interested in improving the protection of buildings from lightning. Engineers had been using copper cables to direct the lightning from the rod to the ground. The lightning rod was invented by Ben Franklin back in the 1700s as a result of his experiments with electricity. It was commonly believed that electricity would follow the path of least resistance when striking a house and because of its low resistance copper was used. However, sometimes the lightning left the wire and entered the structure, causing damage. Lodge thought lightning was a form of DC electricity.

In 1888, in a series of experiments using Leyden jars to simulate lightning, Lodge discovered the existence of electromagnetic waves. In a darkened room he observed a visible glow emitting from the Leyden jars, showing the presence of the waves. Happy with the unexpected result, he decided to go on vacation. Lodge was unaware of the work of Heinrich Hertz and while he was away, Hertz published his findings giving himself credit for the discovery of electromagnetic waves and proving Maxwell correct. While Hertz sent his waves across open space, Lodge sent his over a wire. The two men began to correspond and became friends. Lodge never felt any animosity toward Hertz for getting the credit. In his experiments and demonstrations Hertz built what was essentially a wireless telegraph machine sending dots like Morse code; he did not send voice or sound.

Lodge had also helped refine a device called a coherer, which was a glass tube filled with iron fillings. Originally invented by French scientist Edouard Branly (1844–1940) in 1890, the coherer would become agitated when a current was introduced into it and the fillings would cling together. Lodge discovered in 1893 that it did the same in the presence of electromagnetic waves. The work of both Lodge and Crookes was marred in the eyes of the scientific mainstream because they drifted into metaphysics, coming to believe that the living could converse with the dead. The religion of Spiritualism had begun in the United States in the 1840s. Some, like Lodge and Crookes felt that as electromagnetic waves could be sent over the air and if all energy was just a different manifestation of the same thing, then spirits of the dead might be able to be contacted. They even felt that death itself might be an application of the laws of thermodynamics: a change from one form of energy into another, and not death in the traditionally accepted form. In the end Lodge, like Hertz, was interested only in proving Maxwell correct, not in inventing radio. In fact, Lodge thought wireless telegraphy impracticable as he, and most scientists at the time, thought the waves would behave as light and be able to travel only in straight lines. If they were, then they were ill-suited to long-range communication due to the curvature of the earth.

Nicola Tesla may have built a form of radio transmitter in 1893 but his laboratory was destroyed by fire. He did file for wireless transmission patents in 1897 and 1898. In his work on radio, Tesla was primarily interested in wireless transmission of electrical power with signaling being only one possible application. He was preoccupied with radio-remote control over radio broadcasting, envisioning remote control in a military application. One problem with Tesla and radio was that his genius ranged far and wide, and he tended to be drawn to grand projects. One of his first thoughts after working out the principles of radio, for example, was to plunge into an idea for a worldwide network with impressive relay towers and enormous voltages. Unfortunately, the mercurial nature of his intellect and interests often prevented him from focusing on a single invention or idea long enough to bring it to a practical end. So, while his work on radio-related concepts did predate that of Marconi, he never brought all the pieces together into a working system that could be adopted by others.

GUGLIELMO MARCONI

Of all the claimants to the title of the "Father of Radio" (there are none for "Mother of Radio"), the one most associated with it is the Italian amateur inventor Guglielmo Marconi. While no one person conceived the idea and made all the necessary parts by themselves, Marconi was the one to bring the parts together and promote not only wireless telegraphy, but also himself. He saw the commercial applications of sending wireless telegraphy messages and determined to make it work. In many ways Marconi was to radio what Morse was to the telegraph. Both men were essentially inspired amateurs who used existing technology to realize a dream of creating a worldwide communications system. Marconi (1874–1937) grew up on a private estate outside Bologna, Italy, known as the Villa Griffone. His father was a silk manufacturer and his mother, Annie Jameson, was the daughter of the Jameson whisky family of Ireland. Guglielmo was the younger of two brothers and was raised within an expatriate English community in Italy. As a result, he spoke fluent English and carried himself like an English gentleman.

Though he did not receive a formal technical education, young Guglielmo read technical works in his father's library voraciously. An early favorite was the work of Michael Faraday. He saw himself in the Faraday mold: the self-taught scientist making great discoveries about the nature of the universe. He was also enamored of the life of Ben Franklin, going so far as to try recreating the American's famous kite-flying experiment. When Heinrich Hertz died in 1894, Marconi decided to try to replicate his experiments. A friend of the Marconi family was Italian physicist Augusto Righi, who had performed experiments in electromagnetism himself and often discussed Marconi's work with him. Marconi was so successful in performing Hertz's experiments that he felt he might have discovered waves Hertz had been unaware of. He also noted that Hertzian waves tended to pass through objects instead of bouncing off them and that they could travel great distances.

What Marconi originally wanted to do was run a Morse code printer by using a coherer. The problem was that once the iron fillings of the coherer clung together, it was difficult to get them to separate. As long as the fillings clung together the current would pass through. This was a problem because Morse code required the signal to start and stop rapidly. The tube needed to be shaken after each signal to get the fillings to disperse and be ready for the next signal. To solve the problem Marconi devised a coherer that would shake itself loose after each signal. This way every time he tapped the key, the fillings would cling together, the signal would pass through, and then the coherer would right itself so that the next signal would go out. He built this into a wooden box as his transmitter and in the second box was a doorbell receiver. By the summer of 1895, Marconi had his first working wireless telegraphy machine. He began testing his device outside the Villa Griffone. He discovered that if he attached a wire to the device, he boosted the power and was able to send the signal further. He also saw that if he raised the wire above ground the effect was increased. With his device operational, he headed for London.

On December 14, 1896, Marconi gave his first public demonstration at Toynbee Hall in England. He was introduced by his English patron, the chief electrician for the British Post Office, William Preece. His relationship with Preece gave Marconi entrée into the corridors of power in England and gave him audience to people and institutions the young Italian would have been unlikely to get on his own. He met Preece through his mother's British society contacts. Preece was an enthusiastic proponent of technology and ensured that the post office adopted the latest equipment, including the telephone. He wanted the U.K. post office to lead the way in the use of wireless telegraphy and eventually do away with its conventional telegraphy service. Preece arranged for the hall and put up advertisements. As a result, a considerable crowd showed up to watch.

Marconi brought out his two boxes and set them up. He took the transmitter and began tapping the key. To the audience's amazement, the bell in the receiver box rang each time he hit the key. What astounded the audience even more was that the two boxes were not connected by a wire. Preece took the receiver box in his arms and carried it around the room, allowing members of the audience to look closely at it while showing it was no parlor trick. Anywhere the box was carried, the bell still rang every time Marconi hit the key. During the demonstration Marconi sent no voice or even coded signal; he just rang a bell at a distance using only electromagnetic waves: that was enough for the moment. The evening was a big success. After the Toynbee Hall exhibition, Marconi was interviewed for a number of publications and his experiments were highlighted in the press.

At the time when Marconi was making his first public demonstration, an Indian engineer named Jagadis Chandra Bose (1858–1937) was doing similar experiments. Bose was a Bangladesh-born, British-educated engineer whose method of detecting electromagnetic waves was an improvement on Hertz's original crude design. In 1895 Bose demonstrated the ability to send electromagnetic waves at the Presidency College in Calcutta. While Marconi was giving his demonstration at Toynbee Hall, Bose was in England lecturing: they were both interviewed by *McClure's Magazine* in March of 1897. In 1899 Bose developed a coherer like Oliver Lodge's, but used mercury instead of iron fillings. Marconi employed Bose's more efficient device in a slightly adapted form in 1901. The Marconi camp downplayed Bose and his work, saying he had added nothing to the discussion. For his part, Bose praised Marconi and stated that he was not interested in any commercial application for his work.

Bose was not the only person working on wireless. As far back as the 1860s it was being investigated. A Virginia dentist named Mahlon Loomis (1826–1886) may have managed to send wireless messages from Cohocton Mountain to Beorse Deer Mountain, a distance of 18 miles. He used kites as antennas hooked up to telegraph keys. Like Marconi, years later, Loomis sent only electrical impulses, not true sound, and was issued a patent for his work (US129, 971) in 1872. U.S. Senator Charles Sumner tried to get a government grant for Loomis to continue his experiments, but was only successful in getting the government to allow Loomis to set up the Loomis Aerial Telegraph Company. He kept working, replacing the kites with towers, but the system proved unwieldy and inconsistent. Unfortunately, Loomis was seen as a fraud by many and so his reputation suffered. Unable to find financial support, his work stalled and he faded into obscurity. In 1896, a Russian, Aleksandr Popov, demonstrated a receiver for electromagnetic waves at the St. Petersburg Physical and Chemical Society. Eugene Ducretet in France and Karl Ferdinand Braun of Germany also were doing experiments into wireless telegraphy with varying levels of success.

After the Toynbee Hall demonstration, William Preece arranged for a demonstration of Marconi's boxes before Queen Victoria. Despite some initial problems, it worked and the Queen was duly impressed. From this demonstration Marconi's fame began to grow as did popular interest in wireless telegraphy. At the Queen's demonstration, a German engineering professor named Adolphus Slaby was in attendance—invited by Preece. Following the demonstration Slaby hurried back to Germany and with backing from the German government began replicating Marconi's work. Marconi had something of great potential economic wealth on his hands and so offers to back his work poured in. The young Italian had to be careful; he had, after all, built his machine largely from technology invented and worked out by others. Desperate for what to do, he contacted his businessman father Giuseppe. The elder Marconi had not looked well upon his son's interest in wireless. He thought the whole thing a waste of time and not very profitable. Then, however, his business acumen would come in handy to secure control of the new technology. Family connections between the Marconis and Jamesons soon raised enough capital—over £100,000—to create the first Marconi company, secure patent rights, and pave the way for further research.

To begin the real trials of his device, Marconi established a base of operations in 1897 at the Royal Needles Hotel in Alum Bay on the southern coast of England overlooking the channel. He set up his equipment and erected an antenna and outfitted several ships with receivers that regularly plied the waves nearby. These experiments went well and led to the next major demonstration. The *Dublin Daily Express* asked Marconi to cover the popular boating race, the Kingstown Regatta, for the paper. Happy to oblige, Marconi set up his equipment on a chase ship and sent back word on the course of the race. Besides being a very attention-getting gimmick, the Kingstown Regatta experiment showed that transmission would work off a moving ship and that the signals would pass through heavy rock cliffs. The media coverage of Marconi increased when, following the Regatta, he sent messages across the English channel to France: at over 30 miles, the longest transmission distance to date. More impressive demonstrations were yet to come.

James Gordon Bennett was the publisher of the *New York Herald* and a yachting enthusiast. He had followed with great interest Marconi's work in transmitting news of the Kingstown Regatta and wanted to do something similar in the United States. Coming up in October of 1899 was the prestigious America's Cup race: that year to be held off the New York coast. Bennett offered Marconi the then princely sum of $5,000 to come to New York to broadcast news of the race as it was happening. New York was full of tourists that October as the city had just hosted a major celebration of the U.S. victory in the Spanish-American war. Bennett made a point of advertising that Marconi was going to cover the race. Marconi set himself up, much as he had for the Regatta, in a trailing ship to the racers and sent back constant progress reports. These were the first confirmed public wireless transmissions in America. Much was said about Marconi himself as much as about his device. It was not only the beginning of widespread interest in wireless in America, but also interest in Marconi. He was called modest and unassuming, and in turn he praised America saying that he held great admiration for the country and its people. This only endeared him in the public eye even more. He was being cast by the media in the role of the inventor-hero.

Marconi's success at the America's Cup led not only to great publicity, but also media speculation about the uses for his machine and speculation over the eventual mastery of voice transmissions: something Marconi had not planned on. What can be called the social construction of radio began here. Marconi was about to be christened the "Father of Radio."

Social construction refers to how people build up an image of a technology. This image often has nothing to do with the technology itself, but with how it affects us or how we "see" it. Radio is *radio* because we have built up an image of it in our minds and culture. What we think of as radio, with all the attendant signs, symbols, and meanings we attach to it today, began with Marconi's coverage of the America's Cup and James Gordon Bennett's promotion of it. In 1900, the company he started with his family's money officially became the Marconi Wireless Company. That same year the Marconi International Marine Company and American Marconi also came into being. The Marconi Company and the media embrace of the young Italian from England were so successful at spreading the interest and acceptance of wireless, and of Marconi himself, that the technology came to be associated almost exclusively with him. Wireless transmitters came to be known as Marconi Devices and the messages they sent as Marconigrams.

While the public in England and the United States was amazed by Marconi's demonstrations, the professional engineering world did not see him in quite the same way. The American tradition of the inventor-hero was the inspired tinkerer like Edison. Marconi fit that mold, but the growing professionalism of American science and technology looked down on such men (Edison liked to disparage academically trained scientists as wasting their time on useless detail that had no practical value). The professionals were upset because the media embrace of Marconi made it seem as if wireless had just popped out of his head fully realized, and not that he built upon more than half a century of previous work by others. Marconi was being put forward in the American press as a sort of wizard who conjured radio out of thin air. Trained engineers were not particularly impressed by Marconi's work. They saw many of the inherent flaws in its design. The spark gap that actually generated the electromagnetic wave was slow; the coherer was unreliable and it worked on only one frequency, so more than one transmitter operating at the same time would interfere with each other, and it could not be tuned very well if at all. A major problem also seemed to be the one that plagued the telegraph, distance. It was still thought that wireless had only limited range: about 30–40 miles. This last problem was something Marconi wanted especially to disprove. On November 9, 1899, following his triumph with the America's Cup, Marconi sailed back to England onboard the liner *St. Paul.* On the trip, Marconi set up a wireless set.

He had alerted his team back at the Royal Needles Hotel to be listening for his transmissions. Eventually, the *St. Paul* came within range and the ship and station began exchanging communications.

In the summer of 1900, Marconi began work on his first purpose-built transmitting station. He placed it in the relatively remote region of Cornwall, England. The site, near the Poldhu Hotel, contained a protective fence enclosing a small house and a large antenna array. In addition to distance, Marconi was working on that other great problem of wireless telegraphy, tuning. Electromagnetic waves are measured by the space between the hills of the wave. By changing this distance you change the wavelength. If station X is transmitting on a certain wavelength, or frequency, then receiver Y must adjust, or tune, itself to the same wavelength in order to hear it. With the early spark gap style transmitters, based on Heinrich Hertz's original device, a wave of basically fixed length was produced that allowed anyone with a receiver to hear it. That year Marconi filed for a tuning patent and was granted one numbered 7777.

Using the Poldhu station as his base of operation, Marconi determined to break the distance barrier and show once and for all that wireless telegraphy had practical application for business and especially for the maritime shipping industry. It was a risky move because his fame had reached a considerable level. Any misstep or failure would not only adversely affect his personal reputation, but also that of the fledgling technology. As a result, he began his transatlantic signaling project in secret. The Cornwall area did not attract large crowds and the few reporters who went there to see what Marconi was up to were given cover stories about his experiments. He needed a corresponding site in North America. He initially chose Cape Cod, Massachusetts, but then settled on the more appropriate St. Johns, Newfoundland. At Poldhu, he set up large multi-posted antenna arrays. At the same time, he employed a London University professor named Ambrose Fleming to help build large and powerful transmitting engines necessary to send a signal across the Atlantic. Fleming would later gain fame for his work on vacuum tubes. Marconi then went to St. Johns and began listening for the prearranged signal. In December 1901, he and his assistants George Kemp and Percy Paget waited intently to hear the signal, the single Morse code letter *S*. On the 12th at around 12:30 in the afternoon Marconi heard it. It was an important moment; it showed that, at the very least, wireless telegraphy could send a signal across the Atlantic Ocean, more than 2,100 miles. The signal was faint, but it was there. He tried to keep his success a secret at first until he confirmed it, but word got out: the amateur wizard Marconi had wracked up another impressive triumph.

Guglielmo Marconi with his assistants George Kemp and Percy Paget in St. Johns, Newfoundland, during their efforts to receive a radio signal across the Atlantic Ocean in December 1901. David Sarnoff Library.

Along with the accolades for breaching the Atlantic, the Anglo-American Cable Company threatened to sue Marconi to stop his experiments. They realized that wireless could replace cable transmissions because they would be less expensive to send. While crude, Marconi's success showed that wireless had the range to be a viable communication technology. It seemed so clear that the Canadian government approached him about setting up a permanent transmission station under government auspices. The Canadian post office was interested in beating the Anglo-American company at its own game by offering cheaper communication between North America and England as well as ship-to-shore traffic. A year after this initial success, Marconi was sending complete messages and, by 1907, had regular wireless telegraphy service between Glace Bay, Nova Scotia, and Clifden, Ireland. At the same time, a communication link was set up between Bari, Italy, and Avidari, Montenegro. Marconi was more famous than ever. He was given honorary awards and, in 1909, shared the Noble Prize for Physics with German wireless telegraphy experimenter, Karl Braun.

COMMERCE

In a later chapter, the question "who owns radio?" will be discussed. The start of that argument was in patent disputes that began in the early decades of the twentieth century. Marconi and others began patenting wireless devices and that led to fights over who did what. By the 1920s, radio very quickly became a hot industry with trained engineers and physicists working furiously to outdo Marconi. His companies tried hard to take control of the entire industry but others wanted a piece of the action. A race began to develop more efficient, cheaper, and safer radio equipment.

It was the shipping industry that was first attracted to wireless telegraphy and a number of British, European, and American steamship companies began to equip their vessels with it. In addition to commercial shipping, one of the first major users of wireless telegraphy was the Royal Navy. The British Empire was founded on and maintained by its navy. The ability for warships to communicate with each other and their home base over great distances was obvious. The United States Navy eventually also became interested in wireless and began to invest heavily in its use. In 1898, as a result of the Spanish-American war, the United States then had a little empire of its own: adding Puerto Rico, Cuba, and the Philippines to its overseas acquisition of Hawaii and Alaska. Businesses with offices around the world also saw the advantage of wireless communication as did journalists. By 1905 and 1906, shipboard wireless operation had become fairly common. By the end of the first decade of the twentieth century it was standard equipment.

The world stock market expanded dramatically in the late 1880s and saw the beginning of a mania for corporate merging. The wireless telegraph industry, in its infancy, was able to take advantage of this situation to grow. Marconi and other radio pioneers rushed to create corporate entities in order to dominate the field. A number of forward-looking business attorneys and investors were introduced to Marconi in 1899 following his much publicized successes. They saw wireless as a technology of the future and wanted to exploit it and so formed the Marconi Wireless Telegraph Company of America and opened offices in New Jersey. The organization, eventually known simply as American Marconi, put together a leadership structure heavy in lawyers and politicians in order to create a powerful corporation, which could fight off the inevitable patent lawsuits that were sure to follow. One drawback of all the Marconi companies was the vision of their namesake. While wanting to dominate the field, Guglielmo Marconi saw the market rather narrowly. He wanted his child to grow up at sea as a technology geared to the marine shipping industry. His vision for the future of wireless did not include the child talking. As a result, his empire would flourish only briefly.

The more successful future of wireless and then radio was held by others.

Marconi looked to the oceans as the place wireless would really excel. He saw a network of ships as the prime consumer of the technology. These long-range goals would be undermined if short-term goals of financial success were not met. One problem was that in order for a shipping company to make use of the technology they could not just buy sets and put them on their ships, but had to have corresponding ground stations and trained operators. This meant a major investment most companies were unwilling or unable to make. Marconi proposed that shipping companies could rent or lease the equipment and even the personnel. This allowed the Marconi Company to control the entire operation, earn revenue, and have the network (by leasing Marconi equipment, the shipping line had to agree by contract to not receive or initiate communications with non-Marconi signals). While Marconi may have been trying to dominate the field, others were making contributions of their own.

REGINALD FESSENDEN

As wireless telegraphy spread, a few realized that the next logical step was sending not just dots and dashes, but the human voice. A wireless telephone was needed. The telephone could transmit voices and other sounds, but it still needed a wire like the telegraph. A Canadian émigré to the United States, Reginald Fessenden (1866–1932), like Nikola Tesla, had worked for the Thomas Edison Company and then Edison himself. Also like Tesla, Fessenden's tenure with the "Wizard of Menlo Park" was relatively brief. The Canadian reasoned that if Maxwell had been correct then sound could be sent without the need for a wire.

Born in Quebec, Fessenden was a child prodigy who was awarded a college scholarship in mathematics at the tender age of fourteen. As a child he saw a demonstration of the telephone by Alexander Graham Bell. He left Canada at eighteen, taking a teaching position in Bermuda. In 1885 he was hired to work for Edison and was quickly promoted, but financial problems caused his layoff. He soon found a position with George Westinghouse and worked on the project to light the 1892 Columbian Exhibition in Chicago. He then went to teach electrical engineering first at Purdue University and then the University of Pennsylvania.

As a professor, Fessenden embarked on a program to make the theory of wireless into a reality. Fessenden thought that sound waves radiated out in the way waves would radiate in water away from the spot where a stone was dropped in. From this he hypothesized that high-frequency transmissions would allow for voice, not just clicks and snaps. In 1900, he went to work for the United States Weather Bureau with a special contract which stated that while the Bureau could make use of any of his inventions, he would retain the patent rights. In 1901, he developed the "heterodyne" receiver that could change the high-frequency waves generated in a wireless telegraph's spark gap into low-frequency waves. These low-frequency waves were what made the diaphragm in a telephone receiver vibrate. Realizing that any device transmitting a voice would need to generate a continuous wave, he worked on an alternator to do so. The sparks created by a wireless telegraph transmitter were produced at a fairly slow rate. They would have to be produced very quickly for sound. Fessenden's alternator produced waves quickly and continuously instead of intermittently. In a series of experiments with wireless telegraphy under government auspices, he sent voice transmissions over greater and greater distances. Beyond the voice ability aspect of Fessenden's equipment it also allowed the wave to travel much further than a Marconi machine. He wound up quitting the Weather Bureau when he felt they had reneged on their agreement. He jumped right into the business by starting the National Electric Signaling Company and built a large transmitter at Brant Rock on the Massachusetts shore. He was able to get the United Fruit Company to install his transmitters and receivers onboard their ships that plied the North Atlantic coast. As a publicity stunt on Christmas Eve of 1906, he began transmitting to ships at sea. The wireless operators on duty suddenly heard not the usual Morse code taps, but human voices and music coming through their headphones. Fessenden himself sang and played the violin: he had made the first radio broadcast.

There was a drawback to Fessenden's machine. Though he had succeeded in sending wireless voice transmissions, the signal tended to be weak and erratic, and tuning into it clearly was difficult. As a result, Fessenden had a hard time getting anyone interested in what he was doing. Why should anyone care? The telephone worked well on land and wireless telegraphy for the high seas was more reliable than some scratchy disembodied voice. Another element that plagued Fessenden was that he was not quite the amiable self-promoter Marconi was. He could be curt with people and was often viewed as difficult to get along with. Fessenden's supporters pointed out that when Marconi made his famous transmissions from England to Newfoundland in 1901, all he did was send Morse code in one direction. Fessenden sent true voice first to the ships and then two-way transmissions from Scotland to Massachusetts in 1906. Regardless, Fessenden was unable to interest any backers or make his system pay. He went on to develop other projects related to sound waves. He invented a number of devices for submarine communications and detection, as well as a device for locating icebergs.

THE TITANIC

Fessenden's interest in detecting icebergs was prompted by the most celebrated maritime disaster of all time. The ability of wireless telegraphy to transmit timely information and to serve the public desire for news was highlighted most effectively by the wreck of the RMS *Titanic* on April 15, 1912. The *Titanic* disaster caught the public's imagination for a number of reasons, including the fact that 1,635 passengers and crew lost their lives, and for the role wireless played in the drama. The *Titanic* was built by the firm of Harland and Wolff for the British White Star Line. While famous for its unrivaled level of first-class passenger luxury, the *Titanic's* real source of income was from steerage passengers who could be packed by the hundreds in the lower parts of the ship. It was meant to be a major carrier, not so much of Europe's elite, but of Europe's poor going to America to start new lives as immigrants. It was equipped with the latest model Marconi wireless machines and a pair of specially trained operators. While the builders of the great ship never said so, the media labeled the ship "unsinkable." It was the epitome of the nineteenth-century Industrial Revolution concept of progress through technology and the conquest of nature.

The *Titanic* made only one trip. About three-quarters of the way between the British Isles and New York, it struck an iceberg and sank. Design flaws, including not enough lifeboats for all aboard, doomed the ship. The presence of Marconi equipment on the ship was partially negated by the endless stream of messages being sent out by the wealthy passengers to friends in America. Unlike many ships at the time, which often turned their receivers off at night, the *Titanic's* operators kept them running 24 hours a day. The flood of frivolous transmissions, not unlike the modern phenomenon of people speaking endlessly into their cell phones, tied the sets up and hampered incoming messages about dangerous floating ice.

In the evening of April 15, as the *Titanic* sped along, an enormous iceberg loomed out of the dark. The crew desperately tried to avoid it, but struck the iceberg cutting a long gash down the ship's side under the waterline. Immediately after the collision, the *Titanic's* wireless operators, J. G. Phillips and Harold Bride, began sending out calls for help using the standard CQD and the new SOS (three dots, three dashes, and three dots) distress signals. The SOS signal was first adopted in 1905 by Germany with Britain following suit the next year, though it was slow to be actually used. Several ships were in the vicinity including the *Baltic, Cincinnati, Prinz Albert,* and the *Titanic's* sister *Olympic.* They all heard the distress calls and raced to the scene. The nearest ship that heard the *Titanic's* call was the *Carpathia*, which was traveling from New York to Naples, Italy. The *Carpathia's* captain immediately ordered his ship at best speed to get to the *Titanic* and render assistance. Despite this heroic effort—dangerous because they would be sailing at speed through an ice field—the *Carpathia* did not reach the scene of the wreck until well after the *Titanic* went under.

As ships in the area came to the realization of the scope of the disaster, the air crackled with wireless transmission traffic. In the early hours of the incident all was chaos, few knew exactly what was happening other than the *Titanic* was in trouble. Compounding the problem, as the *Titanic*'s generators went out of commission as it slowly slipped under the water, the electrical load reached critical and caused the Marconi messages to become jumbled and broken. As a result of all the confusion, a number of erroneous messages went out from other ships and were picked up by ground stations which in turn sent them even further. Variously, the *Titanic* was reported as being towed in, that she had hit an iceberg but was proceeding under her own power, and that all aboard were safe. One report had the *Carpathia* picking up all the *Titanic*'s passengers to put on her own decks. Even as late as the next night, the White Star Line's main office was saying that everyone was safe. So many sets broadcasting at the same time on the same airwaves jumbled messages and led to great confusion and erroneous conclusions. Eventually, all messages coming from the *Titanic* ceased. On the *Titanic*, chief wireless operator Phillips kept doggedly sending out position reports well after the captain had ordered all to abandon the ship. At the last minute both Phillips, who was twenty-five years old, and Bride, who was twenty-one, went into the water. Bride was injured but rescued; Phillips perished from the intense cold of the water.

Out of the confusion, the terrible reality began to emerge. The *Titanic*, the largest mechanical object ever built by humans, had hit an iceberg and sunk without a trace, taking most of its passengers and crew with it. In all, only 700 people were saved. The news that the mighty ship was lost went out over the wireless communication system to a stunned world. One of the survivors picked up by the *Carpathia* was Harold Bride. Despite his legs being badly injured, he went to work helping the *Carpathia's* wireless operator, Harold Cottam, send out lists of the living and the dead. A station in New York City picked up these messages and then relayed them out again to the waiting public. The station was operated by an ambitious, twenty-one-year-old Russian immigrant named David Sarnoff. He carefully took down the lists of names and sent them out again. Sarnoff became something of a hero for his efforts and went on to a long, distinguished, and pioneering role in the development of radio.

LEE DE FOREST

The *Titanic* disaster drove home with brutal force the importance of wireless communications not just for routine messages but for breaking news and even lifesaving operations. Despite this, wireless was not yet radio. What was needed was something to make radio viable. In 1901, while Reginald Fessenden was at work on his heterodyne receiver, Lee De Forest (1873–1961) was putting together what he called a responder: an improved signal receiver. Convinced he was on to something, De Forest rounded up some financial backing, primarily from a stockbroker named Abraham White, and opened the De Forest Wireless Telegraph Company. As De Forest's responder grabbed attention, so did the company. White was a shrewd businessman and promoter and the company was soon flush with cash. White and De Forest divided responsibility for the company with White handling business, and De Forest running the laboratories and engineering shops. In 1906, Fessenden sued the De Forest Company for patent infringement and won. White then used the suit as a pretext to fire De Forest.

Forced out of his own company, Lee De Forest resolved to keep working on wireless telegraphy. In the 1890s while working on improving the light bulb, Thomas Edison discovered that electromagnetic waves would pass through a vacuum. This came to be known as the Edison effect. In 1904, an English physicist named John Ambrose Fleming (1849–1945)—who worked for British Marconi—discovered that if alternating current ran through a modified light bulb it changed the AC to DC. Fleming added a pair of electrodes to the device and called it a diode. De Forest became interested in Fleming's device (often called a valve) and read all the relevant literature on it and came up with a modification of his own. In experimenting with a diode, De Forest saw that if a small wire grid were added to it, a sensitive electromagnetic wave detector would be created. The device could pick up faint electromagnetic waves and amplify them so that they could be detected more clearly. As De Forest had added a third element to the diode it seemed logical to call it a triode. De Forest, however, argued that as his invention could pick up the continuous waves necessary for audio broadcasting, it should be called an audion. There was the piece of the puzzle to make human voice broadcasting reliable. De Forest had invented the vacuum tube: the backbone of radio and later television broadcasting.

De Forest believed there would be a mass market for wireless communication, but it was years before he could move ahead with his plans. He was stymied by a series of legal entanglements and financial problems. Through all this, he tried to improve the sound of broadcasts so they would

Lee De Forest holds one of his audions, the three-element vacuum tube that made broadcast radio a practical technology for the mass market. Courtesy David Sarnoff Library.

The vacuum tube developed after World War I became a standard for home radios, showing its cathode and grid, the plate anode, and the glass bulb silvered by the effort to increase the vacuum. David Sarnoff Library.

be of a quality that people would listen to. On the verge of bankruptcy, he sold his audion patents to AT&T (the company J. P. Morgan began with Edison), which was expanding its telephone business and felt the audion would improve phone-voice quality. As much of an improvement the audion was, it still did not quite work as well as needed. While it could detect weak signals, it did so in a limited fashion forcing operators to employ headphones to hear it. As De Forest became entangled in legal issues, he did little to advance his invention. Some argued that De Forest did not really understand the nature of the device he had created. To use a radio metaphor, the audion needed some tuning. The audion was brought to its full potential by a Columbia University electrical engineering student named Edwin Armstrong. De Forest felt that it was DC flowing through the vacuum tube. Armstrong thought it was AC. In 1912, his investigations showed that it was indeed AC. This was an important discovery as AC can be manipulated and tuned, whereas DC cannot.

Born in New York, Edwin H. Armstrong (1890–1954) was inspired as a teenager by the life of Marconi to begin investigating wireless telegraphy and other aspects of electrical engineering. While in high school he began building his own devices including a 125-foot-tall antenna in his yard. He purchased many commercially available electrical devices including some De Forest audions. While at Columbia, he discovered that if he manipulated the inner workings of an audion tube, he could cause the current to feedback upon itself, amplifying the signal. By pumping up the feedback the signal picked up was made so loud that headphones were not needed. The improved audion could act both as a receiver and an amplifier thus allowing wireless telegraphy to finally become radio. Armstrong's work brought him to the attention of the U.S. government and he entered the Army Signal Corps as an officer when the country entered World War I. He was sent to Paris to work on building a system to pick up enemy wireless signals. He managed to build a device using several improved audion tubes that pulled in signals louder and more clear than ever before. He called the device a "super heterodyne" circuit. Instead of starting his own company in the 1920s, Armstrong sold some of his patents to Westinghouse and the German firm Telefunken (a result of Germany's early embrace of Marconi technology), and others to the newly formed Radio Corporation of America.

EARLY REGULATION

By 1912, wireless telegraphy traffic had risen to a level where jams were beginning to occur. The relatively narrow frequency range generated by the primitive sets quickly filled to capacity. The United States Navy had come to depend on wireless communications; government, corporate, as well as commercial stations were broadcasting. Added to this traffic was an increasing number of individual amateurs sending messages. That year, the U.S. government passed the Radio Listening Act (RLA). Less a rule about listening than broadcasting, the RLA required a license to send out signals, whether by a corporation or an individual. It also required operators to have passed a special test in order to be issued the license. The law also gave Congress the authority to issue specific frequencies to licensees as well as call letters for identification and time slots for broadcasting. The airwave dial was broken up into three chunks with a large one for commercial broadcasting, one for government purposes, and a small section—at the extreme short end of the spectrum—for individuals. Finally, the law gave Congress the right to levy fines for inappropriate behavior like using too much power or crossing over into someone else's frequency. Interestingly, there was little in the way of regulation of content (the exception being sending out false emergency signals). The government justified this control by claiming that the airwaves belonged to the public and in order to ensure fair use of that air, the government would keep a watchful eye over it.

The Radio Listening Act also allowed the government to seize control of privately owned equipment in case of an emergency. When World War I broke out in 1914, the United States (though still neutral) expected to be drawn in sooner or later, and government and military planners knew wireless would play an important role. Therefore, they wanted to have as much radio equipment on hand as possible. In order to get manufacturing into high gear, the government put a moratorium on the seemingly bottomless pit of patent rights lawsuits. They allowed any company that had the technical capability to do so to start manufacturing radio equipment regardless of whether it had the patents. When the United States did enter the war, the government cracked down on the amateurs and ordered their sets shut off for the duration. Some of the larger commercial stations—mostly American Marconi stations—were taken over outright by the navy. U.S. government support of foreign-owned firms like Marconi began to turn away and move toward American-owned companies. Contracts went out to fill huge orders of radio equipment needed for the war. Big companies like General Electric (GE), AT&T, and Westinghouse turned over parts of their operations to the manufacture of wireless equipment for the government. The war created a boom in radio-related manufacturing. It also ended the brief honeymoon period when the airwaves were genuinely free for the use of anyone. After the war, the battle to control the airwaves became increasingly intense. Supporters of the RLA argued that it was the government's takeover of the wireless industry that helped win the war. Opponents shot back that it was ironic that a war to end autocracy in Europe would result in the creation of a broadcasting autocracy in America.

After World War I, a scramble to control the radio industry ensued. There already existed a number of communications companies by the end of the nineteenth century. Western Union and AT&T controlled the conventional telegraph industry, while GE, Western Electric, and Westinghouse were major producers of electrical equipment and generators. A series of acquisitions and mergers formed the nucleus of what became the powerhouses of the industry. These companies were then poised to make enormous profits and greatly influence radio. GE was created in 1892 when the Edison and the Thomas Houston companies merged. This brought together under one roof, power equipment, electric motors, and a significant number of patents for electric trains and trolleys. Western Electric controlled a large chunk of the telephone industry along with crucial patents. AT&T then purchased Western Electric.

Along with Marconi, other inventors started their own companies. Lee De Forest started working for Western Electric in 1899. Due to his hard work and ambition, he was quickly promoted to the telephone research department. He then went to work for the American Wireless Telegraph Company where he invented his responder. The company wanted the device but De Forest refused to hand it over and so was fired. In the following years, De Forest started a number of companies in order to build and market his inventions. In 1901, he helped found De Forest, Smythe, and Freeman. Part of De Forest's problem was that, like Tesla, his mind was brilliant but eccentric; he would come up with one good idea after another, but his interest would wane before he could make a real business out of it. In just two years between 1901 and 1903, he put together four companies, none of which were successful. In 1910, another De Forest company, the Radio Telephone Company, broadcast the first live opera performance starring Enrico Caruso. The company went bankrupt the next year. He then formed the Radio Telephone and Telegraph Company and began manufacturing audions. The Marconi Company immediately sued, saying the audion design infringed upon the Fleming Valve that Marconi owned. This resulted in audions not being available for research and amateur radio construction. The De Forest Company could barely hold its own against Marconi, but then AT&T was granted a licensing agreement to make audions by Marconi. The writing was on the wall for De Forest. The De Forest Company was run out of business by the early 1930s and what was left was taken over by a new corporate creation.

During the war, the U.S. government confiscated the British-owned Marconi Company in America. When the war was over, Congress pondered what to do with it. A plan was worked out by the Secretary of the Navy, the chief of naval communications, and Owen Young, the president of GE, to transform the American Marconi Company assets into a new and fully American company. Its operations would be controlled by Americans, its stock owned by Americans, and the original foreign owners reduced to a mere presence. With little they could do to stop it, American Marconi acquiesced and in 1919 gave up control and the Radio Corporation of America (RCA) was formed. Along with a virtual monopoly on ship-to-shore wireless broadcasting and manufacturing in America, RCA also quickly acquired the patent rights from Lee De Forest and Edwin Armstrong for vacuum tubes. As American Marconi had acquired the patents to Ambrose Fleming's diode, RCA got that too. They were also able to buy the patent rights Armstrong had sold to German Telefunken. Young became chairman of the board with Edwin Nally as president and a young executive named David Sarnoff as commercial manager. RCA then signed an agreement with AT&T in July of 1920 allowing AT&T to acquire a considerable block of RCA stock, and arranged that all the radio and radio-related patents both companies owned would be available to the other for a decade, royalty free. In that year, Westinghouse merged with the International Radio and Telegraph Company (formally known as NESCO). This gave Westinghouse control over Reginald Fessenden's heterodyne patents. Westinghouse then attached itself to RCA with the agreement that RCA would buy 40 percent of its equipment from Westinghouse and the rest from GE. In 1926 with all the merging, cross licensing, benefit agreements, and sweetheart deals complete, RCA, GE, and Westinghouse formed the first entertainment and news broadcasting network: the National Broadcasting Company (NBC). In 1929, RCA acquired the Victor Talking Machine Company and its ability to produce commercial recordings. RCA could now control radio manufacturing, radio broadcasting, and the production of music to play over the network as well as the music being recorded. The RCA consortium had done the same thing the defeated European powers of World War I had tried to do: carve out an empire of struggling colonies. Instead of an empire of land, RCA had created (in author Tom Lewis's words) an *Empire of the Air* (1991). It was a powerful example of technological empire building.

CONCLUSION

By the 1920s, what had been just a dream in the 1890s had come to be. Radio had been born and had taken its first tentative steps. There was still work to do, but it was off to a good start. Some still ask, however, "who invented radio?" In some ways, who gave birth to a technology is less important than who raised it. Nikola Tesla may have done the former, but not the latter. With the telegraph, Samuel Morse received the lion's share of the credit as it was he who put forward the system that was ultimately adopted, even though many worked on the development of the technology. Guglielmo Marconi propagated wireless telegraphy beyond an intellectual exercise and showed the world its importance. That Marconi received the credit as the "Father of radio" should in no way diminish the contribution of Tesla, De Forest, Fessenden, and others. It is a sad reality in the history of science and technology that sometimes crucial figures are forgotten or given short shrift. In the end, while Tesla described radio and its possibilities, Marconi built it and sold it. Partisans argue that Tesla was publishing papers on wireless communication by 1893 and that Marconi simply took Tesla's work and capitalized on it. Marconi claimed to have never read any of Tesla's work, but only that of Hertz. By 1897, Tesla had received his first patent on radio communication and the next year built and operated his remote-controlled boats in New York. In what was something of a paternity suit in 1943, the U.S. Supreme Court acknowledged Tesla as having invented radio, arguing that Marconi's patents were nothing Tesla had not already done. Regardless of who gave birth to radio, others would now take over to raise the child.

3

Plastic and Transistors

In the study of the history of science and technology, it is often assumed that science comes first and then technology. A working definition of science is that it is a search for knowledge of the nature of the universe. Technology is an application of knowledge to some practical, mechanical end. In other words, a discovery is made of some knowledge (science) and then applied to a working device (technology). While this is often the case, it is not always. Sometimes a device or process is stumbled upon before the inventor understands why it works the way it does. In the twentieth century, a number of inventions and scientific discoveries assisted in the growth of radio. Not all of them were designed with radio in mind, but benefited the technology anyway. An understanding of the nature of the atom helped lead to the invention of the transistor, which allowed for a phenomenal growth in the electronics industry. The transistor in turn led to the invention of the integrated chip, which not only led to the invention of smaller radios but also to the modern computer. Of the directly related radio breakthroughs, frequency modulation (FM) may have been the most significant. FM allowed for a much better sound quality. This helped spur a growth in more complex popular music. Radio was a part of and at times a leader in technological advance.

RADIO AND THE NAVY

An early influence on the life of radio was the world naval community. The British Royal Navy was the first to embrace the new technology. They were quickly followed by the Germans and the Italians. The United States Navy was slower to get into the radio business. Just how enthusiastically the navy adopted wireless and then radio is a point of some debate. The navy had a long history and a tradition of adopting new technology slowly. It had adopted steam power, armor plating, and revolving gun turrets only grudgingly in the face of the Civil War. After the war, the navy had been left in large part to rot. Tradition-bound officers were reluctant to adopt any new technology they did not fully understand. Toward the end of the century, media reports on the navy's condition—and that condition's threat to national security—prompted the government to go on a major rebuilding program. After 1898 it was clear to jingoistic expansionists inside the government that America's future rested on the back of a large navy, which could project American power around the world. That navy would also have to be able to communicate with base stations and other ships at sea over great distances.

The American navy was a warren of bureaucracies that guarded their political niches jealously. Adopting such an important technology like wireless would be far easier said than done. In 1899 Marconi put on some demonstrations in which United States Navy electrical experts observed and reported back favorably to their superiors. Later that year, the warships *New York* and *Massachusetts* were equipped with sets for testing. The tests were successful, but the navy observers saw the same inherent problems other engineers had. They eventually turned Marconi down because they felt he was too expensive (he demanded a royalty as well). They also thought his rule about not contacting non-Marconi networks a bit much. In addition, the hierarchy of the United States Navy was predominantly ethnically Anglo-Saxon and Nordic and viewed the Irish-Italian, Catholic Marconi as a suspicious foreigner. Marconi thought that the navy was being heavy-handed and unappreciative of the technical aspects and difficulties involved. The navy then tested wireless sets from most of the larger companies of Europe as well as some American firms. The navy required ease of use and simple maintenance even at the sacrifice of distance. The companies the navy went to were unwilling or unable to supply demands they thought strange. The first order contract went to the German firm of Slaby-Arco. Reginald Fessenden was incensed that the navy ignored him and refused to buy his sets. It eventually began to use Marconi equipment, but only after a decade of negotiations. Many were miffed at Marconi's demands for exclusivity and were prepared to do something about it. The Marconi Company steadfastly refused to answer any transmission from non-Marconi equipment. This tactic was growing infuriating to many. The German government called for an international congress on wireless for 1903. Just before that, Slaby-Arco and another German firm merged to form Telefunken. The British thought the call for a congress a purely political move. The primary topic of the congress was the Marconi monopoly. While not accomplishing much, the congress did show the American navy just how far behind everyone else it had fallen and drove it to renewed effort. As a result, American wireless telegraphy was whipped into shape and attaining goals such as coast-to-coast transmitting capability and more shipboard equipment were accelerated. The navy was put in charge of all government wireless operations. To some, it smacked of a government monopoly replacing the Marconi monopoly. Instead of buying a system, the navy built its own, a system that soon showed itself to be below commercial quality. With low-quality equipment and officers most of whom were at first halfhearted in their interest in wireless, the navy blundered on. Eventually, the navy came around and placed great emphasis on wireless and radio and began operator schools. It became a leader in the use of the technology.

The navy was not the only organization which saw the need for trained operators to work the technology correctly. In 1909 the Marconi Company realized that in order to spread the use of its equipment, it needed to begin an institution to train people. The Marconi Institute was opened in New York as a place to train young men in the use of this revolutionary technology. They learned the basics of electromagnetic wave behavior, Morse code, and other technical knowledge necessary to operate wireless telegraphy. When American Marconi was dissolved into RCA, the school was renamed as the RCA Institute and became a premier school for the training of radio and television engineers. In the 1970s, the RCA Institute was separated from the company and it became the TCI College of Technology and is still in operation.

CRYSTAL SETS

One thing that helped spread interest in wireless telegraphy among average people was that if a person could not afford to build a full-scale set, there was a low cost version they could turn to. The alternative was the crystal set, which was a crude device capable of picking up radio waves: something far easier than generating them. In order to pick up radio waves, a set had to be

The Marconi Institute, New York, 1912. This was the first school to teach wireless telegraphy. Courtesy of the Library of Congress.

able to change a high frequency wave to something less powerful, which then could be heard over a telephone receiver. The process was known as rectifying. In order to do this properly and effectively, researchers like Fleming and De Forest devised complex vacuum tubes. For wireless to work on a commercial level, increasingly sophisticated equipment was being developed. This cut out chances for amateur or consumer involvement. In 1906, however, an army general named H.H.C. Dunwoody discovered that the compound of carbon and silicon, called carborundum, would work as a simple receiver element. A small wireless equipment manufacturer patented the process, and the crystal set was born. The crystal was placed between a pair of copper contacts and then a thin wire (called a cat's whisker) was manipulated over it to achieve a simple form of tuning. It picked up both Morse code and later voice transmissions.

One of the other hurdles to get over was the need to be able to tune wireless so that specific signals could be picked out and not confused. An early attempt to solve the problem was the work of John Stone (1869–1943). A trained engineer who had built a successful career at Bell Telephone, he applied for a tuning patent in 1900. Originally working with the standard setup of a spark gap transmitter and coherer, Stone soon found that it was this very arrangement which was causing the problem. The transmitter was putting out a clear signal but the coherer often failed to pick it up or at least pick it up in a fashion which made it discernable from the static. He rebuilt the set so that the receiver and the coherer were one (he also used carbon instead of iron inside the coherer). He eventually hit upon an electrolytic detector that worked much better. A series of tests with other scientists showed that Stone's device was able to pick out a particular wavelength and not just the strongest local signal. Despite interest from the United States Navy, Stone and his company were unable to gain the type of attention necessary to make his tuner a success. Eventually, better tuning was worked out, but that did nothing to alleviate the narrow range of sound quality of radio.

FREQUENCY MODULATION (FM)

The Fleming Valve and De Forest's audion were the inventions that allowed for the creation of a steady carrier wave needed for voice transmission. The original Hertz spark gap produced only impulses, which created the dots and dashes of Morse code. Tube technology supplied the continuous wave and allowed for the addition of a speaker to the receiver so that cumbersome earphones were not needed. This allowed for individuals to listen without being physically tied to the set and also allowed for groups to listen at the same time. The dominant medium of radio was amplitude modulation (AM). Amplitude modulation is where the power of the signal (amplitude) is modulated or varied and the frequency (the carrier wave) of the signal remains constant. AM radio worked by sending signals out at specific frequencies. If a station's call number was 1190AM then the listener would turn the dial on their receiver to 1190 to be able to listen to that station. The number means that the station is sending out its signal 1,190 times a second. The sound is attached to the carrier wave by adjusting the amplitude of the wave. The station's main amplifier then pumps the signal out at a certain level of power (usually 50,000 watts). The waves then radiate (hence radio) out from the station into the atmosphere to be picked up by any receiver tuned to 1190AM. The signal is then converted back into audible sound by the listener's receiver. This system opened up the process to static as any power source being generated near the receiver would interfere with the signal's reception. There were many who were unsatisfied with AM's abilities and disappointed by its limitations. In the late 1920s Edwin Armstrong, the man who invented the improved audion or super heterodyne circuit (which made radio broadcasting a viable issue), was already looking for an alternative. Armstrong's answer to the problem was to modulate not the amplitude, but the frequency. By keeping the power consistent, the static problem was eliminated. With the frequency modulated and the carrier wave amplification constant, FM was born. In the 1930s Armstrong began a series of experiments and field tests, which showed that FM had a startlingly clear sound and was able to reproduce a wide range of subtle tones in high fidelity. Unfortunately, FM was born amidst the heights of the Great Depression. This kept radio manufacturers from adopting an entirely new system when they were putting all their efforts into getting people to buy AM sets. There was neither the money nor the interest in adopting a new system to do something already being done. Armstrong took his idea to representatives at RCA who were impressed by its abilities, but they had a problem. RCA, like the rest of the radio industry, was already wedded to AM. They were also in the early stages of research into television and were investing heavily in that new technology. Taking on another technology at that point was considered impractical (also, both FM and television worked in the same VHF range so they would compete for broadcast space).

Being rebuffed by RCA, Armstrong took a chance on developing it by himself. He went to the Federal Communications Commission (FCC) for a broadcast permit and it opened up a part of the spectrum between 42 and 50 MHz for FM broadcasting. It was not until 1940 that Armstrong was able to put up his first FM station at Alpine, New Jersey, along the Hudson River. The FCC still dragged its heels for another two years before granting him a frequency to broadcast. The outbreak of World War II pushed FM into the background, and after the war FM and television began to battle over airspace. It was solved by changing the frequency allotment for FM to the higher end of the scale at 88–108 MHz. Superficially, this was a good thing as more space for stations was created; however, the shift made the existing FM equipment unusable. As a result, FM did not really catch on for another 20 years. One of the things that helped keep FM afloat in the early days was the peculiar concept of "store casting." This was a special pay service on FM frequencies geared especially for stores, medical offices, and banks. The broadcasts of music were uninterrupted by commercials or host voices and were intended to supply such institutions with a steady stream of soothing background music. Another thing which helped boost FM's popularity was that in 1961 the FCC allowed stations to begin broadcasting in stereo: the use of a multiplex, or double signal, to break up the parts of the transmission onto separate channels (AM was a mono signal). After the war, FM broadcasting did start to expand, but Armstrong himself was dogged by corporate challenges to ownership of the system. By the early 1950s Armstrong was spent economically, intellectually, and emotionally. In despair, he committed suicide in New York on January 31, 1954. His wife continued the fight and eventually won a series of law suites, which awarded her millions of dollars in infringement settlements. Despite the sad end to the life of Edwin Armstrong, FM went on to explode as the medium of choice for a new generation of radio listeners who were interested in the sound quality as much as the content of radio.

ATOMIC THEORY

Separate from radio research at first, but having a later impact on it, was work on the structure of the atom. Ancient Greek philosophers first used the term atom to describe what they thought would be the smallest building block of the universe. This term was resurrected by John Dalton in 1803 when he proposed an atomic theory of the nature of chemicals and electricity. As more was learned about the structure of chemicals and the arrangement of atoms in them, the Russian Dmitry Mendeleyev (1834–1907) began arranging chemicals in a chart in what became known as the Periodic Table of the Elements. A few years later it was theorized that electricity was made up of negatively charged particles called electrons. Using a device first invented in the 1850s called a cathode ray tube, William Roentgen (1845–1923) discovered in 1895 that certain chemicals glowed when exposed to the tube. The glow was in the form of energy waves that were not deflected by magnets. He called these unknown emanations X-rays. The next year Henri Becquerel (1852–1908) saw that the effect was a result of the spontaneous decomposition of certain elements. The Polish chemist Marie Curie (1867–1934) and her husband labeled the spontaneous decay process, radioactivity. More discoveries about the nature of atoms were made until the early 1920s when physicist Niels Bohr (1885–1962) proposed that the structure of the atom looked like a sun with little planets orbiting it in discrete layers called shells. Each shell had a specific number of electrons in it while the central mass, the nucleus, was a combination of neutrons and protons.

The Bohr model of the atom helped explain a long-standing problem of electricity: why did some materials conduct electricity and some insulate from it? Scientists long knew that materials like wood and glass would not conduct electricity, while metals would. It was hypothesized that the arrangement of electrons in the outer shell of an atom gave it its conducting or insulating qualities. Atoms in which the outer shell had a full complement of electrons would not allow electricity to pass through it: this made it an insulator. Other atoms, while they might be able to hold a certain number of electrons in the outer shell, did not. This open space in its outer layer allowed electricity to pass through the atom making it a conductor.

Some elements, however, had the ability to throw off outer shell electrons, making space where there was none or pulling in electrons to fill spaces in the outer shell. This changing condition allowed the atom to alternately conduct and insulate, making them semiconductors. Semiconductors appear naturally in the form of germanium, zinc, and silicon (which is formed from beach sand). Silicon is particularly good for electrical applications. When a pure form of silicon is mixed with other materials—known as dopants—the resulting material acquired a high degree of semiconductibility. This condition opened a new area of research. If a semiconductor could be controlled, a number of possibilities suddenly became possible.

TRANSISTORS

The original Fleming Valves were unreliable because of the inability of engineers to produce a lasting vacuum inside the glass bulb. In 1913, the American Irving Langmuir (1881–1957) designed a way to get a better vacuum. Some French military engineers used Langmuir's process to make improved triode tubes for military applications. A number of companies followed suit and began producing large numbers of these better tubes. With World War I over, these new tubes went on the commercial market and inventors, engineers, and amateurs gobbled them up.

While they proved a major boost to the technology, by the late 1930s, radio had gone about as far as it could go with vacuum tubes. The tubes did good service but still had limitations and drawbacks. The vacuum tube began its life as a light bulb that produced both light and heat. Vacuum tubes produced a marginal amount of light, but they did get hot. They took a moment to begin to operate once they were switched on. This warm-up period delayed the action of the radio and generated a lot of unwanted heat. They burned out easily and were susceptible to vibration. Miniaturized vacuum tubes cut down on the problem, but an alternative was needed. That alternative was the transistor (from the combination of the words transfer and resistor). The transistor was essentially a vacuum tube without the extraneous glass and vacuum. A tiny triode, the transistor's three elements, allowed for the amplification of electrical signals. The nature of the transistor also made it an electrical switch. By combining a number of transistors together, a wide range of electrical devices (including radio) were possible.

The pioneering work on transistor technology began at the Bell Telephone Laboratories in New Jersey in the mid-1940s. A group of scientists and engineers were working on the nature of crystals. This research was begun in 1938 by Russell Ohl and seen by co-worker Walter Brattain (1902–1987) who thought it could prove to be the basis of an alternative to the vacuum tube. Brattain was born in China but raised in Washington. He received a Ph.D. in physics in 1929, and immediately went to work for Bell, specializing in the study of semiconductors. Originally working with copper, he switched over to germanium as a source material. He was joined by John Bardeen (1908–1991), who was a Ph.D. in mathematical physics. Bardeen and Brattain discovered that by putting two contacts on a crystal they had a device, which operated much as a vacuum tube did. Both men realized that one of the problems they were facing in the crystal research was the impurity of the materials they were working with. They were trying to see how the surfaces of the crystals behaved toward current but were not getting very far. Bell began experimenting with crystals when it was discovered some time earlier that certain crystals would act as a rectifier, allowing current to pass through them. During the war, Bell used this discovery as a central part of its work on radar. At Bell, the semiconducting element germanium was recognized, along with silicon, as a good material for this purpose. Germanium and silicon were relatively simple crystals that contained a natural lattice-like structure similar to the wire grid of a Fleming Valve. In 1947, team leader William Shockley (1910–1989) had been working on this but ran into difficulty with it; so he turned it over to Bardeen and Brattain and then left them on their own. Both of Shockley's parents were engineers and he was born in London. After his education at Cal Tech and MIT, he went to work at Bell. During World War II, he led an antisubmarine research team and consulted for the U.S. Secretary of War. Following hostilities, he found himself leader of solid-state research at Bell with Bardeen and Brattain. They used germanium as a semiconductor but could not get any result either. At one point Brattain submerged the model they were working on in water and it began to work a little. Bardeen finally cracked the problem by hypothesizing that electrons in the crystals formed a barrier at the material's surface. They then built a strange contraption of razor blades and paper clips, which acted as an audio amplifier. On December 16, 1947, the Point Contact Transistor became a working reality. One of the advantages of the transistor was that it worked immediately after it was turned on, needing no warm-up period. The two men did not tell anyone what they had discovered at first, but kept working on it. They then went to Shockley with their revolutionary invention. Shockley was happy for the success but upset that the two men had not kept him apprised of what they were doing. Not wanting to be left out of the discovery, Shockley took the point contact transistor away from them, reconfigured it over the course of a few days, and came up with the Junction Transistor. He used several slices of semiconducting silicon because he found it would allow current to be both rectified and amplified. Shockley's device was more reliable than the contact transistor and easier to mass produce. After a brief period of working out the kinks, a patent was applied for in early 1948 and a public announcement of the invention of this revolutionary technology was made in June of that year. The success of the transistor with Shockley's name at the top annoyed Bardeen and Brattain, making them feel like they were being squeezed out. What was once a close-knit team of researchers began to break up although all three men were awarded the Nobel Prize for their work. Shockley quit Bell and started his own company in Palo Alto, California. Because of his personal manners, a group of his best scientists quit his company and formed their own. Shockley ended up losing his company and then ruining his reputation by publicly supporting the idea that human intelligence was linked to race. The transistor itself did not make a big splash initially. Most companies saw it only in a military role. Transistors were eventually adopted for a wide range of devices, creating an explosion in the electronics industry. Transistors could be attached to boards along with resistors and diodes to make complex switching, amplifying, and transmitting devices. The most popular transistor-based product was the portable radio.

TRANSISTOR RADIOS

Almost immediately after the announcement of the transistor a number of people began thinking of ways to use it for commercial purposes. Up to that time radios had been fairly large, cumbersome devices, filled with hot vacuum tubes, housed in ornate wooden boxes and plugged into a wall outlet. Radios for household use took the form of furniture. By the 1920s, there were already a few "portable" radios being produced. The "Operadio" appeared in 1923 and was the brainchild of the father and son team of J. McWilliams Stone, Sr. and Jr. While technically portable, the Operadio was about the size of a mailbox with a carry handle on top and was touted as "Radio Wherever You Go!" Hit hard by the Great Depression, the company turned to making loudspeakers for military and then commercial purposes. The dream of a portable commercial radio persisted, however. Vacuum tube technology had begun to produce smaller versions, which allowed for battery-operated, man-portable radios for military use during World War II. Instead of wooden boxes, military field radios came in metal boxes or sometimes a material called Bakelite. Plastic eventually became the material of choice for electronic devices beginning in the 1950s as it was easy to mass-produce, could be made into any shape or color, was heat resistant, and did not conduct electricity. The first plastic (the word comes from having the ability to be molded into different shapes) appeared in 1862. Created by Alexander Parkes, his Parkesine was an organic compound made from cellulose, but its high cost prohibited it from being used for commercial purposes. Celluloid was developed in 1866 as a replacement for ivory in the making of billiard balls. It found use as the base for photographic film. The first genuinely synthetic plastic was Leo Baekeland's liquid resin. Introduced in 1907, Bakelite had all the positive characteristics of modern plastic and was used for a variety of applications particularly in the military. It was resistant to electricity and acid degradation and it did not fade in color from overexposure to the sun or corrosion from salt water. Bakelite was still on the expensive side, however. Advances in plastics technology continued throughout the war years. A DuPont chemist named W. H. Carothers came upon a polymer substance by manipulating the chemical chain making up the material. As a result, nylon, saran (the basis of Saran Wrap), and teflon were discovered. The most important of these, for the history of radio, was polyethylene. This new plastic had all the right properties but was also inexpensive to produce and easy to mold into any shape. It was ideal for mass-producing portable transistor radios.

The first commercially operational "pocket" radio which made use of both transistor and plastic technology was the American made Regency TR-1 released in October of 1954 (possibly the first genuine transistor radio was made by Texas Instruments the previous May, but not mass-produced). Although several types of transistor radio appeared in Germany and the United States, the most influential was the TR-63 of 1957 made by a Japanese company. A little over a month after World War II ended, Japanese engineer Masaru Ibuka founded a company to build transistor-based products for the consumer market. Ibuka began the tiny electronics firm in the ruins of Tokyo, Japan. Originally called Tokyo Tsushin Kenkyujo, the company was chiefly involved in repairing existing radios. Later that year Ibuka was joined by an old colleague, Akio Morita. In an effort to appeal to a wider and more Western market, the company began using the name Sony. It was an important step as few other Japanese companies used the Latin alphabet for their logos.

In 1954 Sony received the go-ahead to use Bell Laboratory's patents and Western Electric's license to manufacture transistors in Japan. By the next year it had produced its first pocket-sized radio, the TR-55. Sony went on to produce a line of radios that were sold mostly to Europe and Canada. By 1957 Sony was producing the breakthrough model TR-63. It came in a colorful box with a leather carry case, a 9-volt battery for power,

The combination of transistors and plastic revolutionized the electronics industry and helped spread radio use further than ever before. Harcourt Index.

an earphone, and an instructional booklet. The transistor radio appeared at just the right moment to be caught up in a cultural phenomenon. In the 1950s and 1960s, rock 'n' roll began to boom first in America, then England and Europe, and then worldwide. The portability of the plastic-encased transistor radio appealed to the legions of young people who were the music's natural fans. The transistor radio became standard equipment for many teenagers and led to a revolution in portable communications technology that continues to this day.

By 1958 the transistor was already reaching its physical limitations. Researchers at Texas Instruments and Fairchild Camera (later Fairchild Semiconductor) began experimenting by putting several transistors on the same piece of semiconducting material. One of the problems of the transistor, a holdover problem from vacuum tubes, was that each transistor performed only one function. Transistor radios had to give over space to diodes and capacitors, as well as transistors. They also had to be hand wired, which slowed production. The Fairchild project was an effort to find a more efficient way to mass produce semiconductors. In 1959 the group, including engineers Robert Noyce and Gordon Moore, unveiled a process using very thin sections of silicon, which allowed the connections to evaporate themselves into the surface, eliminating the need for hand wiring. Known as the Planar process, it also eliminated the need for the metal capsule of the transistor. It allowed the device to be wired to do more than one function: combining transistors, diodes, and others parts into one object. The result of this approach was the invention of the Integrated Chip (IC). This new device went on to replace the transistor as the transistor had replaced the vacuum tube. The small size of the IC allowed for more operations to be performed and led to the development of the modern computer. Noyce and Moore left Fairchild to form Intel Corporation.

CONCLUSION

The mostly technical part of the life of radio will now give way to the part that was dominated by the life of radio as a medium for communication and culture. A good deal of philosophical discussion has occurred over what role technology, and especially communication technology like radio, has had on human history. It must be remembered that despite the fact that many refer to radio as if it were a living thing—radio does this, radio does that—it is a machine. How it affected the lives of millions of people around the world is not the result of radio doing something on its own, but on how it was used by people. That idea will take center stage for the rest of this story.

4

Private to Public

By the end of World War I, radio had gone from a temperamental and experimental technology to a reliable and practical one. The interwar years (1919–1939) were a period of important growth for radio. It went from a private form of communication to a public one and a major cultural icon on top of it. As this was happening, a role reversal took place: radio was a product of culture, now culture was being created by radio. As a mass-market medium, radio acquired the ability to not just transmit information and entertainment, but to be a molder and image maker for the societies that embraced it. Radio would set the pace for behavior, belief, fashion, and popular culture, and people listened and followed. Every society on earth had trendsetters and cultural pioneers, whom others imitated and used as inspirations for tailoring their own lives. Radio was able to do this on an unprecedented scale. Through radio, local culture became national culture, which in turn became world culture. While the earth had been steadily shrinking for centuries, radio accelerated the process to a dizzying speed. Users and producers of radio soon realized its power and strove to capture it. Like an orphaned child of great beauty, power, and potential, many looked enviously at radio and schemed to control the course of its life.

DAVID SARNOFF

The *Titanic* disaster of 1912 affected the lives of all aboard. One person not on the ship was also affected and that was David Sarnoff (1891–1971). He was a twenty-one-year-old wireless operator at the Wanamaker's Department Store in New York. He came on duty shortly after the ship went down and spent hours sending out the names of the lost and became a hero for it. He was able to parlay this celebrity into a major career as a radio pioneer. Born in Russia to impoverished Jewish parents, he was the eldest of five siblings. His father came alone to America and worked to earn money to bring the rest of the family over in 1900; but they lived in poverty. Seeing his father work so hard for so little in return, young David vowed to become successful. He went to work for the Commercial Cable Company, which was an American subsidiary of the British company that operated the transatlantic telegraph cable. His tenure there was short but it was at Commercial Cable that he was introduced to the telegraph.

His imagination fired by the telegraph, Sarnoff found work with American Marconi. His rise was swift and he found himself as a personal messenger to Marconi himself when the boss was in New York. He volunteered for wireless training and was posted to a desolate corner of Cape Cod, Massachusetts. In 1905 he was promoted as night manager at the Marconi station on Coney Island, New York. After working at several East Coast wireless telegraphy stations, Sarnoff spent time as a shipboard operator on a seal hunting ship. Following this nautical adventure, Sarnoff became the Marconi operator for the station at Wanamaker's Department Store. It was from this post that Sarnoff found fame for his role in the *Titanic* disaster. The Marconi Company was impressed enough by his performance that night that they promoted him.

After the *Titanic*, Sarnoff worked his way steadily up the ladder at American Marconi. Sarnoff's political acumen was considerable and making him even more formidable was that he had a dream and a vision for the future of radio. While a number of people were already thinking of the future with wireless communication as a mass medium, Sarnoff was in a position to actually do something about it. In 1916 he suggested an idea to the vice president of American Marconi, E. J. Nally. He saw a role for a wireless-based entertainment system where the listener had only to listen. He envisioned a radio music box that would bring entertainment, news, and music into the home. The owner of the receiving device need not be an engineer or mechanical tinkerer. All they would have to do would be to switch the device on, tune to a station by using a simple dial, and enjoy what came out. His other suggestion was that the company should start

David Sarnoff sits before National Broadcasting Company microphones on his desk on the 53rd floor of Rockefeller Center's Radio Corporation of America Building, 1936. David Sarnoff Library.

manufacturing simple, inexpensive radio receivers geared specifically to this consumer market. Sarnoff 's persistence paid off. When Radio Corporation of America (RCA) was created out of American Marconi, he transitioned smoothly to the new company and in a couple of years was RCA's commercial manager initially, and later general manager.

There is speculation that Sarnoff 's memo to E. J. Nally may have been apocryphal: that only after he had begun his plan to create a radio entertainment network did he create a story of prophesying it in an earlier stroke of genius. The "Radio Music Box" memo, as it has come to be known, on its face is quite historic as it lays out a vision of the future that actually happened. However, David Sarnoff was a tireless promoter of David Sarnoff and was accused of occasionally taking other people's ideas and putting them forward as his own. Some of the initial legend of the memo came from author Gleason Archer's *History of Radio to 1926* (1938). He knew Sarnoff personally and supposedly had access to RCA archives. Some argue that Sarnoff was inspired in his vision by Lee De Forest's station 2XG, which broadcast in New York during 1915–1916. De Forest was putting out a music-oriented show received by other radio enthusiasts and ships in the New York area.

De Forest's station, which he called "Highbridge," had announced the presidential election results of 1916. The *New York Herald* was spreading word of the voting by telephone to various public places around the town, including De Forest's radio station. He claimed that he personally broadcast the news over the air. An early radio enthusiast magazine said that De Forest was trying to create a newspaper sent by wireless. David Sarnoff was aware of De Forest's efforts and may have seen the future in it.

Once at RCA Sarnoff proposed, in a 1920 memo to his boss Alfred Goldsmith, the need for manufacturing portable radios for consumer sales and creating an entertainment network that consumers could tune into. In that memo, Sarnoff referred to his supposed Radio Music Box memo of 1916. He certainly wrote the 1920 memo, but the existence of the 1916 memo is in question. The 1920 memo did outline a vision of the future of broadcasting. Regardless of whether David Sarnoff saw the future of radio in 1916, he certainly saw it in 1920 and worked ruthlessly to bring that vision to reality. Alongside Sarnoff there were other radio enthusiasts with visions of their own.

THE AMATEURS

Wireless telegraphy seemed a natural outlet for nonprofessionals interested in new technology. After all, had not Marconi himself been an inspired amateur? He became a hero to legions of (mostly) young men around the world, but particularly in the United States. A number of publications appeared, geared to the growing amateur movement. Instructions for building wireless sets began to appear in mainstream technology-oriented magazines like *Scientific American* and *Popular Science*. The first two decades of the twentieth century were a period of freewheeling amateur activity with sets appearing all over and operators trying to contact each other by Morse code and listen in on others.

An early amateur celebrity was twenty-six-year-old Walter Willenborg of Hoboken, New Jersey. A student at the nearby Stevens Institute of Technology, Willenborg built a station at his home and was listening in on Marconi's transmissions between Nova Scotia and Ireland. The *New York Times* ran an exuberant story on the engineering student's work and labeled him a "boy hero." The youth-oriented magazines of the day promptly adopted amateur wireless tinkering as not just an area young people could have fun with, but at the same time acquire the scientific and technological knowhow, which would be necessary for the coming century and to lead them to righteous adulthood instead of slovenly idleness. An image was being created in America: an image of the proper young person who did not waste their time with vain pursuits, but engaged in hobbies where they learned skills and behaviors that as adults would lead to successful lives, self-sufficient and manly (girls were not generally a target for this kind of promotion). The whole notion of amateur wireless telegraphy fit well with the American ideal of taking charge and leading the way for a country that was rapidly becoming a world power. Male inventor-heroes sprung up all over American popular culture. Tom Swift, Nick Carter, and others (Jonny Quest was a later twentieth-century version) were young men who relied, not on punching fists like earlier heroes, but on sharp minds. These stories told of incredible adventures open only to boys with inquisitive minds and technical skills. These boys did not wait for someone else to do it for them and they did not rely on luck; they actively made their own futures with a "can-do" attitude. Wireless operation would be a necessary part of the education of such boys. The late nineteenth and early twentieth centuries are considered a golden age of invention in America. Every day seemed to unveil some new mechanical wonder. Electricity was making possible not just lighting homes but lighting a way into the future. It was another phase of the Industrial Revolution and what boy did not want to be part of that revolution? Wireless equipment, both ready-made and homemade, and little in the way of government regulation (at least early on) allowed amateur wireless activity to take off in America in a way it did not elsewhere around the world. Despite the reality that the corporate world of radio was headed nowhere with many companies going bankrupt and the U.S. military's initial ambivalence to radio, the amateurs kept it alive in their basements and garages.

In 1904 the Electro Importing Company of New York began selling not just parts for wireless sets but entire units ready to be assembled with instructions. Begun by a teenage immigrant named Hugo Gernsback (1884–1967), the company sold transmitters under the brand name Telimco Wireless Telegraphy and were first advertised in *Scientific American* in November of 1905. Gernsback became a major advocate, supporter, and promoter of amateur wireless telegraphy issues and worked to forge enthusiasts into a cohesive community. He published a series of magazines including *Modern Electrics, Electrical Experimenter*, and *Radio Amateur News*. His interest in technology, publishing, and the future led him to become an influential publisher of science fiction as well.

Born in Luxembourg, Gernsback received an education in electrical engineering before coming to America in 1904. He had designed a dry cell battery and hoped to market it in the United States, but his success was mixed and he turned to wireless telegraphy. His Telimco sets sold for $7.50

Two boys from Sterling High School in Sterling, Colorado, peer into the future. One wears earphones and twists the knobs of the radio; the second boy operates a transmitter, 1941. Denver Public Library, Western History Collection, Call No. X-10524.

each and proved far more popular than his batteries. In an effort to broaden his market and sell more wireless sets, he entered the publishing world in 1908 in order to promote the use and purchase of his kits. Where Marconi, for example, had seen big companies as the natural market for wireless telegraphy, Gernsback correctly gauged that individuals, particularly boys, would buy scaled-down versions. He also saw a connection between technology and literature. One of the things that drove Gernsback to enter the world of "science fiction" (a term not in use then) was that he felt science-based novels were too fanciful and seemed to have no basis in reality and that the authors rarely explained the technology and grandiose science they placed in the text. He felt that literature based on science and technology should be as realistic as possible and that if stories did not explain science competently, it would not inspire readers to join in and become involved in real science. For Gernsback, the science in the fiction had to make sense. As a result, he began to publish what he called "scientifiction" stories in his radio magazine. This proved so popular that he began *Amazing Stories* magazine in 1926 as a venue exclusively geared to this realistic form of speculative writing. He then dropped scientifiction for the more melodic term science fiction. This word had been used before but was not well-known; Gernsback's use of it brought it into the mainstream.

It was a natural step for Gernsback to go from radio to science fiction. Wireless telegraphy and then radio represented a quantum leap in technology

Though the focus of radio training was on boys, girls were encouraged to participate as performers. Here activist Kathleen Wilson of Arizona oversees a radio program written and put on by children of the Junior Artists Club, 1935. Courtesy of the Franklin D. Roosevelt Library Digital Archives.

and not just for communications. The manipulation of electromagnetic waves opened the road to television, radar, and other new forms and suggested an endless vista of technological change. As wireless telegraphy was a technology that almost anyone could become involved in, many dreamed that radio waves could carry not just the human race in general, but themselves in particular, to the stars. And once out amongst the heavens, there was no telling what wonders would be found there. Science fiction can trace its origin to the late nineteenth century with such works as French author Jules Verne's *From the Earth to the Moon* (1865) and Briton H. G. Wells's *War of the Worlds* (1898), and even as far back as 1813 to *Frankenstein*. One recurring theme in many of these stories was the use of electricity as a medium by which the future could be reached. By the 1920s no self-respecting science fiction story could be without elaborate radio communications devices, radio controlled machines, and radio wave-based death rays. The coupling of radio technology and a greater understanding of the structure of the atom opened the field and helped fictional characters go places they never had before. Generations of young men and women entered the real world of science and technology because of the science fiction stories they read in dime novels and comic books.

By 1910 the amateur community had grown quite a bit, thanks to the efforts of Gernsback and others as well as a general interest in wireless telegraphy. There were hundreds of operators in New York and just as many in Chicago. It was becoming a popular pastime to build and operate one's own set. After 1912 that community became increasingly concerned with government regulation, which it saw as restrictive. While most of the amateurs were law-abiding and well-behaved, some used their sets in ways the government found problematic. As the problem of tuning was still being worked out, it was fairly easy for an amateur to eavesdrop on government transmissions and even send messages masquerading as official naval traffic. In some cases, amateurs had better equipment than the navy. Some amateurs used their equipment to assist the navy when the government's transmitters encountered technical difficulties or broke down. The Radio Listening Act (RLA) was, in part, an effort to restrict the amateurs and force them down the dial and get them out of the government's hair. Gernsback felt that many amateurs were getting involved and that the freewheeling nature of the community was leading to anarchy; and so he publicly advocated government regulation of the airwaves. It was partly his work that brought about the RLA. He also took credit for suggesting that the amateurs be restricted to the lower bands of the radio spectrum as not to clash with government and commercial broadcasting. A shrewd businessman, when the amateurs were driven off the air during World War I drying up the market for his Telimco sets, Gernsback simply repackaged them as electronic experiment kits for kids and continued earning money.

It was around this time that the curious term "ham" came into use. Not all the amateurs possessed the same skill level. It took considerable practice to master Morse code: that the operator had to hold each keystroke longer for wireless than for conventional telegraphy made it more difficult. Some became quite expert while others were sloppy at best. These less skilled operators were sometimes called "hogs" because their slow, bumbling code sending could monopolize the air. These operators were also known as "hams" (like a poor stage performer was known as a ham actor). Eventually, the term ham came to denote any amateur radio operator regardless of their skill level. As they organized, some of the hams tried to repair their reputations by providing communications services in times of crisis as a type of volunteer wireless emergency squad. They would send word of impending storms and flooding and would send information on local events to newspapers and public offices. This led to an attempt to link the East and

West Coasts of America by amateur wireless. Begun in Passaic, New Jersey, with the United Amateur Radio Club, the attempt was made to bring various independent operators and clubs into an early type of Internet. They finally succeeded though it never became a widespread service. Despite these efforts to prove their worth and seriousness, the U.S. government cracked down on the amateurs at the country's entry into World War I. The Radio Listening Act had in it a clause that allowed the government the option of suspending amateur operators in times of war. In April of 1917 the government exercised that option and shut all the amateurs down. (The Canadian government had done the same thing in August 1914.) The U.S. Navy also attempted to ban not just wireless transmitting, but listening as well.

The radio hobbyists were not particularly pleased with the RLA of 1912. It pushed them into a very tight corner away from mainstream broadcasting. As broadcasting had yet to reach widespread public use, First Amendment issues concerned few but the amateurs. The government countered by pointing out that they were not restricting content, only location on the dial. The radio hobbyists were mostly men and boys who through their tinkering with wireless sets sought not just to investigate a new technology and have fun, but to take control of it instead of being swamped under by it. It was common for the young men who populated the ranks of the early telegraph industry to feel themselves part of a special brotherhood, separate from the rest of the populace: after all they were initiates into a kind of secret society who had special knowledge and skills and used a difficult code system that few outsiders understood. The amateur wireless operators too felt themselves part of a special group on the cutting edge of the dawning of a new world. Yet at the same time, some could see that the role of the lone amateur had only a short life. The growing corporate nature of radio would push out and marginalize both literally and figuratively any who did not join up with it. Between 1906 and 1912 the term wireless telegraphy became simply wireless, then radiotelegraphy, sometimes radiotelephony, and then finally just radio (the name stemming from the way the electromagnetic waves radiated out from the source in all directions).

After commercial radio broadcasting became widespread, the amateur movement in America underwent a change. Prior to 1920 or so the amateur operators felt themselves part of a larger enterprise to spread the use of a futuristic new technology. Even though they were engaged in a superficially private activity, closed off from the world by their technical knowledge and headphones, they were building a larger public community. After the 1920s the nature of amateur radio became more genuinely solitary.

Elmer Bucher, sales manager for the Radio Corporation of America, demonstrates his home radio transmitter and receiver, ca. 1921. Courtesy David Sarnoff Library.

They left the cutting edge—big corporations had taken over there—and went into the shadows. Amateur radio became a way for its operators to escape the ordinariness of their otherwise dull lives. Ham radio operation became a way out of isolation, but in reality it only connected them with other lonely kindred spirits. They spent long weeks building their sets in the basement alone like their inventor-heroes Edison or Marconi had done (only the hams were inventing something that had already been invented). Amateur radio ceased being an exploration of vistas yet unseen and became another hobby like gardening or model railroading. After building their sets, the hams then spent their time trying to contact other hams around the world. Indeed, collecting lists of contacts in other parts of the globe became a primary pursuit of the amateurs. The new generation of hams were older than those early pioneers, though still mostly male, and mostly Americans. These operators were increasingly marginalized. Whereas manufacturers sold parts for amateurs to build their individual sets, after the 1920s they increasingly sold completed outfits for listeners, not broadcasters. While the wireless amateurs had looked to a wide-ranging public hi-tech future, the radio hams had looked for a private space where they could create an alternative world for themselves. Taking on aspects of a recreation that promoted gender distinction, the ham radio operators' "shack" became an oasis. Just as the author Virginia Woolf argued that women needed a "room of one's own" as a place to develop their personal character in private, the ham radio operators' shack (basement, attic, or garage) was these men's room of their own; a place they could go to in order to recapture and reinforce their masculinity. This separated the hams even further from the mainstream. Even though they had drifted so far off on their own, or maybe because of it, during World War II, the government became afraid that these private operators might pose a threat as spies and so shut them down again. During the 1950s the Republican Senator Joseph McCarthy, in the midst of his infamous "un-American activities" hearings, called for a permanent shutdown of the hams because they might be sending secret information to the Soviets. By the later part of the twentieth century, ham radio operation was the last genuinely private part of radio. (The late twentieth century would see a resurgence in shortwave ham radio broadcasting as a venue for isolated political groups, mostly right-wing separatists and neo-Nazis to send out their messages to others.) Secure and alone in their shacks, the hams protected their privacy and their masculinity by contacting only those like-minded individuals doing the same thing. Mainstream radio had headed in another direction. The private aspect of radio began to disappear. The private was becoming public.

COMMERCIAL BROADCASTING BEGINS

Not only did the post–World War I period see a boom in the manufacturing of radio equipment, but it also saw the beginning of the commercial broadcasting age. People across Europe, the British Isles, and America were building their own radio stations in their homes, garages, and places of work (some as strong as 4 or 5 watts) and started broadcasting news, music, talk, and whatever took their fancy. These early public broadcasters assumed that others like them would be out there listening. There were a few attempts at commercial broadcasting besides Lee De Forest's Highbridge. In 1916 the American Radio Research Corporation set up an experimental station, call letters 1XE, at Tufts College. Using De Forest audions, the station put out music for whoever might be able to hear it. At first, listeners were unsure of where the music was coming from and a bit of a stir was caused by the mystery broadcasting. The station had its numbers changed several times over the years (the X denoted experimental) but by the mid-1920s was off the air. The most famous of these early efforts was a transmission station set up by the chief of radio production at Westinghouse, Frank Conrad. Building the station in his Pittsburgh, Pennsylvania, garage, Conrad was able to put together an unusually sophisticated transmitter as he had access to the latest Westinghouse technology. At the same time, the Joseph Horne Department Store in downtown Pittsburgh built a receiving station on its showroom floor as a gimmick to advertise their sale of hobby radio sets. To everyone's surprise, they began to pick up Conrad's broadcasts of music. Light bulbs went off in the heads of Joseph Horne's advertising men and they immediately began to encourage people to buy sets so they could listen in on Conrad and other local broadcasters. News spread quickly and executives at Westinghouse asked Conrad to set up a station at corporate headquarters so they could put out election results. The success of the scheme lent support to the idea that people would buy receivers if there was something out there to receive.

Conrad's Westinghouse headquarters transmitter broadcast on a powerful 100-watt system. As per the rules of the RLA of 1912, he petitioned the Department of Commerce (DOC) for a frequency and the station was issued the call letters KDKA. With that, the first of what would be known as a "news flash" announced that Warren G. Harding had defeated James Cox for president of the United States. The November 2, 1920, stunt was so successful that Westinghouse decided to continue on a regular basis. Westinghouse quickly put up another station in its factory at Newark, New Jersey, and another one in Chicago. Following Westinghouse's lead, commercial stations began popping up all over America. The connection between entertainment and advertising became apparent. By 1922 the government was issuing over twenty licenses a month for new stations.

Before commercial radio broadcasting started in earnest, an alternative system was already being tried. The telephone was commonplace by the 1890s and so a number of ventures were attempted to bring a pay news and music service to telephone owners. In 1889 the American author and philosopher, Edward Bellamy, wrote his futuristic vision *Looking Backward* and envisioned a telephone-based entertainment system. The next year, a system was put up in Paris in which coin-operated machines were set up at public places, particularly theaters. They were essentially telephone juke boxes: a patron would put in a coin, put on the headphones, and listen to music. They were more a gimmick than anything else, made to entertain patrons waiting to go into the theater. Two years later, what was the most successful of the telephone-based entertainment systems began in

Budapest, Hungary. The brainchild of inventor Tivadar Puskás, Telefon Hirmondó gave news reports, music, and literary readings to paying customers over their telephones. It was so successful, it was still running in the 1920s. An English phone-based entertainment system was begun in 1895 with subscribers paying a fee for the service (Queen Victoria reportedly found it quite entertaining). Such phone-music systems never really caught on in the United States. In 1911 the Telephone Herald began operations in Newark, New Jersey. It proved popular amongst listeners but the system was short-lived. A few other such programs began around the country but none achieved success.

THE RADIO BOOM

There is some disagreement over what was the first real "radio station." Frank Conrad's KDKA is commonly pointed to though various partisan sides argue for others. What is clear is that beginning in 1920 and 1921, commercial broadcasting stations began appearing across America and that they became popular quickly. As a result of this, the U.S. government began to more strictly define what types of broadcasting would be allowed and by whom. It all had to do with licensing, fees, and control of a technology that from the government's point of view was spinning wildly out of control. Originally, all one needed to get on the air was the equipment. Later licenses were required. There were "commercial" licenses and "standard broadcast" as well as "limited broadcast" categories. These were issued by the Bureau of Navigation (remember, radio's original guardian was the navy). Here the term commercial meant broadcasting specifically for business purposes, particularly the marine shipping industry. When KDKA was issued its license, its call letters were of a type normally given to marine broadcasting stations even though Westinghouse never had any intention of using its station for that purpose. In 1921 the DOC briefly acted as the issuing agency of "broadcast service" permits that allowed for what is thought of today as commercial radio operation: news, music, and entertainment. It was not until the next year that the DOC listed these new entertainment stations separately from their marine cousins. Also, in 1922, the government prohibited the hams with only amateur-class permits to broadcast entertainment content.

Originally, all commercial entertainment stations had to broadcast on a wavelength of 360 meters or 833 kilohertz (kHz). The single wavelength filled quickly and caused problems of overcrowding. Stations were forced

Kolb's bakery company of Philadelphia is the first to install radio equipment on delivery trucks. Note the aerials on the roof of the truck. Date photographed: April 24, 1922. Copyright Bettmann/CORBIS.

into an awkward time-sharing arrangement with each station wanting the best times. The government then opened up broadcasting on 750 kHz. This too proved inadequate and more wavelengths were opened up. This led to a hierarchy of stations on Class A and Class B, then Class C wavelengths. Eventually, a whole range of space on the dial from 550 to 1,500 kHz were made available. In 1926 the DOC was relieved of its radio license granting duties and chaos reigned as permits were given out to almost anyone who wanted one. This led to problems of interference with more powerful stations stepping on the signals of their neighbors. In an attempt to bring order to the chaos, the U.S. Congress formed the Federal Radio Commission (FRC) to oversee the radio industry. Stations then had to have at least 20 kHz of space between them. Throughout 1927 and 1928, the FRC tried to arrange the broadcast dial so that as many stations as possible could use the airwaves, yet allowing those stations clear air to broadcast on. By 1928 the dial looked as it does today. The broadcast day ran from six in the morning until midnight with each time-sharing station having 4½ hours a day. In times of financial trouble, especially during the Great Depression of the 1930s, larger, more financially secure stations bought out their smaller time-sharing neighbors.

BUSINESS HISTORY

Once corporations became involved in broadcasting, they naturally looked to see how they could make profits from it. The beginning of profitable radio can be traced to New York when the AT&T station there, WEAF, began "toll broadcasting." This was the idea that a corporation or other business would pay the station to run its advertising. Initially, this involved a sponsor having its name attached to some specific program and incorporated into the name of the show. The first such venture was the *EverReady Hour*: a variety show sponsored by the battery makers. Another pioneering program at the time was the *Roxy and His Gang* show of 1923. Roxy was Samuel L. Rothafel, the manager of the local Capitol Theater. The show consisted of live broadcasts from the theater featuring various acts with Roxy as emcee. This setup was a lucrative deal for both the station and the theater. The sponsor received advertising and the station made money. To make the deal even better, there was no need to pay the performers extra as they worked for the theater. The format of the show paralleled that of the theater, where the host talked to the live and listening audience and introduced the acts. The host would tell the listening audience what they had just heard or were about to hear. This made the novel idea of radio entertainment familiar, approachable, and less radical to listeners. It also helped establish the idea of a reliable show, that is, a show which came on at the same time every week containing content the listener was familiar with, so that the listener would return to the station over and over (and hopefully buy the products being advertised). It was simply a vaudeville show brought into people's homes. The transition went smoothly and Roxy became the first local radio "personality."

Other stations began to work out reciprocal agreements with local orchestras and operas, like the Capitol and WEAF were doing. When the theater season ended for the year, the station would bring in individual acts to perform or just create its own orchestra that would operate year-round. The first paid regular guest performer may have been the xylophone and ukulele playing singer Wendall Hall. In 1922 Chicago's KYW, a Westinghouse station, put Hall on the payroll. A natural character, Hall used monologues and humor to augment the music. Though raised in Chicago, Hall affected a Southern dialect planting him firmly in the black-face minstrel show tradition, which was still widely popular. This helped make him seem more "down-home" and acceptable to rural audiences. Hall capitalized on his popularity by composing the song, "It Ain't Gonna Rain No Mo' " eventually recording it. He then went on tour doing the show from different stations, local theaters, and music stores. As his sponsor was EverReady Batteries, he toured to promote them as well. Hall was a good fit for the company as its batteries had a distinctive red painted top and he had a head of bright red hair. He became so associated with the product to the listening audience that he became known as EverReady Red. Where Roxy had been a local celebrity, EverReady Red Hall became the first well-known national radio personality. Following EverReady's success, corporations and local businesses began to sponsor shows or even entire stations as they saw the kind of profits radio advertising could produce.

It was not just corporate sponsorship which began at this time but the coupling of radio and journalism came into its own. In 1920 the *Detroit News* became a sponsor and contributor to station WWJ. Following Detroit's lead, nearby Chicago became a center of newspaper-related radio. The most influential of these was the *Chicago Tribune's* World's Greatest Network (WGN). The *Tribune's* first foray into radio was in 1922 when it associated itself with station KYW by providing news segments. Unimpressed with the results, the *Tribune* pulled out. By the end of 1923, however, the *Tribune's* leadership reconsidered the idea. It became convinced that radio was on the edge of a major cultural breakout and decided it wanted to be there. The paper bought into the idea that radio was a virgin child with no one looking out for it. Radio could be exploited by social forces in America that would use it to spread un-American ideas, particularly socialism and women's and racial equality, which would undermine democracy and lead to revolution. Who better, the paper thought, to lead the country away from such dangerous tendencies than the *Chicago Tribune* with its traditions, high moral standards, and good moral judgment? In March of 1924 the self-styled WGN went on the air. The station formed from a merger of WDAP and WJAZ, both popular entertainment stations. As self-appointed protector of America's soul, WGN's content leaned heavily to "serious" music and "high" culture. The paper held the station to a high standard of broadcast quality and news dissemination.

With the spread of commercial broadcasting, the industry began looking to listener markets and audience to see who was listening and how best to approach them. They wanted to get people not to just buy radios and listen in and then buy what was being advertised, but to believe that it was essential for them. One of the first markets to be zeroed in on was the farmer. Radio was presented to farmers as doing several things: it brought useful information to them in the form of stock market news, weather, and various agricultural reports, and it relieved the boredom of isolated farm life. Across America in the early part of the century, rural and isolated farming communities lived spare and repetitive lives. After working all day in the fields, they returned home to supper and two or three hours of leisure time. While this time might be filled by the occasional dance, town meeting, or visitors, there was little to break the monotony. What better way to end this excruciating boredom than to listen to the radio? The introduction of the radio into farm life as a utilitarian and entertainment device changed farm life in ways both radical and subtle.

Part of the underlying message radio sent to the farm was to set up the farm in opposition to the city. All the way back to the earliest days of the republic, there was an intrinsic duality between town and country. The farm was held up as the paragon of all that was good about the new country and the very notion of democracy. The city was where all those things like honesty, courage, and moral fiber were drowned. The radio attempted a reversal of this or at least a rehabilitation of the city. The urban was where everything smart, sophisticated, cosmopolitan, and forward thinking was located. The farm was backward, rude, and isolated. The thing that could cure this problem was the radio (this was going on at the same time some radio programming was stressing the rural over the urban). The radio could bring the city to the country and make it less forbidding and more attractive to country folk. The radio industry, as well as some in the government, felt that farm people were in serious need of being "uplifted" and made more like city people. As a sedentary population, farm people could be targeted as consumers of radio-advertised goods and ideas.

The radio, which had been socially constructed, was then itself constructing parts of society. Country folk were characterized as the "other" in opposition to urban dwellers. In this way, radio technology usage was put forward as a way to be hip and sophisticated and to open parochial eyes to the wider cosmos. If you didn't listen to the radio, you were out of touch both literally and figuratively. Technology was establishing who was cool and who was not. Radio was put up as a necessity and signifier of modern life. This was an important step in the homogenization of America (and the world as well). Mass communication technology brought people together and at the same time tried to get them to conform to a unified lifestyle and thought process. Modernity was what every citizen should strive for, and radio was one of the crucial outward signs of that modernity. "Radio quickly became an ambassador of urban values, and those values became a standard by which to judge rural life" (Patnode 2003, 299). Farmers were also thought to be the one segment of society that could practically benefit from radio. Crop and weather reports helped farmers pursue their business. City people listened to radio for entertainment, but radio helped farmers to be more prosperous. Prosperous American farmers meant a prosperous America.

THE NETWORKS

As explained before, the National Broadcasting Company (NBC) was created in 1926 by Westinghouse and AT&T out of the creation of RCA. The first NBC operations center was opened on 5th Avenue in New York City and began broadcasting in October of 1927. NBC quickly incorporated station WJZ, the Westinghouse station in Newark, New Jersey, set up after the success of its Pittsburgh operation. The next station brought into the network was AT&T's Manhattan station WEAF. By the end of the 1930s they needed new quarters and so moved to the more spacious and opulent 30 Rockefeller Center. "Thirty Rock" became NBC's world headquarters for its radio and later television operations. The NBC facility designed by engineer O. B. Hanson was impressive. It eventually held over twenty studios as well as numerous offices and was packed with cutting-edge broadcast technology. Hanson had attended the RCA Institute and worked his way to chief engineer at the Newark station. He saw that better sound quality resulted by soundproofing the studios and separating performers, musicians, announcers, and technicians into their own booths.

Around the same time NBC was forming, the Columbia Phonograph Company began thinking about radio. In 1927 it aligned itself with a company called United Independent Broadcasters Inc. United would pay a number of local radio stations $500 a week for airtime, which would feature Columbia music. The scheme did not work well and it was abandoned. The music section backed out and United Independent took complete control of the operation and renamed itself the Columbia Broadcasting System (CBS). Businessman William Paley (1901–1990) with the support of his wealthy family acquired a majority of CBS's stock and took control to run the company according to his vision. Originally based out of WOR in Newark, the network quickly brought together the original sixteen stations from the first venture and reorganized and began to make a profit. In the running battle to see which network could out highbrow the other, CBS created its own orchestra and had singers from the prestigious Metropolitan Opera of New York perform on the first day of broadcasting in September 1927. Paley was to CBS what David Sarnoff was to NBC: a powerful personality whose vision of what the network should be permeated every corner of the operation.

Paley was the son of a wealthy cigar manufacturer. He attended the prestigious Wharton School at the University of Pennsylvania, after which he entered the family business. In 1925, while most of the company leadership was away, he took the bold step of spending $50 to sponsor a show on Philadelphia radio station WCAU. Upon learning this, his uncle angrily canceled the arrangement, forcing the show off the air. Listeners complained of the loss and it was seen that cigar sales had gone up because of the advertising. The Paley family rethought the decision and the Congress Cigar Company became a major radio sponsor and saw their sales soar. His appetite was whetted for radio; Paley was happy when the family acquired the faltering CBS radio network in 1927. He quickly involved himself in it and was made president. Paley radically altered the way radio was being used as an advertising tool by giving participating stations content for free instead of charging them for it, which was normal. He also went after the tobacco industry as a source of advertising. As he was familiar with this area of business from his work with the family's cigar company, Paley pursued George Roy Hill, a tobacco industry leader and convinced him to go with CBS. Unlike other networks, CBS allowed sponsors to create their own advertising. This approach was so successful that NBC gave in and did the same. Paley also changed radio journalism by insisting CBS newscasters and reporters stick to factual reporting and refrain from opinion giving. They were to report and analyze, not tell people what or how to think about it. Paley wanted CBS to be an objective and trustworthy news source and so began a new department at CBS just to establish and enforce standards and practices for news broadcasting. In 1946 the New York flagship station of CBS changed its call sign to WCBS. In 1995 it was purchased by RCA's original owners, Westinghouse. The next year, CBS Radio was grafted onto another recent Westinghouse acquisition, Infinity Broadcasting. In 1997 Westinghouse renamed itself the CBS Corporation, and was in turn acquired by the megacorporation, Viacom.

The third of the "Big Three" American radio networks appeared a decade-and-a-half after the debut of CBS. By 1937 the newly formed Federal Communications Commission (FCC) viewed the growth of NBC with alarm and began proceedings against it. In an attempt to preempt the government's ruling, NBC's parent RCA sold off the smaller, less profitable Blue Network to Edward Noble of the Lifesaver Candy Company in the mid-1940s. There had already been an American Broadcasting Company in Seattle, Washington, which went defunct at the start of the Depression in 1929. The new and independent Blue Network bought the rights to the Seattle company's name and the modern ABC network was born. As the decades went by, the Big Three networks got bigger. They expanded from radio to movies, book publishing, news gathering, various entertainment ventures, and, of course, television. A fourth network, the Mutual Broadcasting System, appeared in 1934 out of the smaller Quality Network. Comprising four stations originally from New York, Chicago, Detroit, and Cincinnati, Mutual produced its own content and shared it amongst the four stations. In 1935 Mutual began adding stations in a bid to compete with NBC. While not as well-known today as the other networks, Mutual stood its ground against the Big Three. Unlike the others, however, Mutual never diversified into areas like television and movies. While by 1997 the network had 950 stations, its days were numbered. In 1985 it was purchased by the up-and-coming radio network Westwood One. By 1999 Mutual had gone out of business in everything but name.

During this same period of explosive radio growth in America, other parts of the world were joining in. Transmitters went up in Paris, France; Beijing and Shanghai, China; Cuba; Germany; Belgium; Finland; Norway; Mexico; Australia; Austria; Japan; and Korea. The most influential non–U.S. radio network went up in England. The first sound station in England was the experimental 2MT set up by Guglielmo Marconi in 1920 at Chelmsford. 2MT put out short programs each day of music and talk, which were picked up by listeners employing crystal sets. Seeing the growing popularity of radio both in Britain and the United States, Marconi finally came around to the idea of commercial voice broadcasting. In May of 1922 the British Marconi Company along with a number of other radio-related companies came together. The General Electric Company, Metropolitan Vickers, British Thompson-Houston, and others joined with Marconi to form the British Broadcasting Company (BBC). Unlike RCA which was, despite the involvement of the U.S. Navy, a private concern, the BBC was partially taxpayer funded. The British government was as concerned with Marconi's domination of the industry as was the U.S. government, the Germans, and others, so was reluctant to allow for the creation of the BBC. It gave in, however, in October of 1922 and issued the company a license. In short order, the BBC opened a number of transmitting stations at London, Manchester, and other sites around the country. The BBC operators prided themselves on a high quality of broadcast content. The level of programming brought in more listeners, which spurred the company to greater efforts, which in turn resulted in more radio sales and greater listenership. In 1924 the BBC switched over to long-wave transmissions that extended its reach to cover the entire country. It also began to experiment with stereo sound using long waves for one channel and medium waves for the other.

In 1926 the BBC's original government charter expired. Arguments were made that the network should be made a public trust, not a private commercial enterprise, and so the British government stepped in and took control, or nationalized it, and renamed it the British Broadcasting Corporation (BBC). As a concession to the original spirit of the network, while it was technically a government operation, it was granted a Royal Charter which allowed it to be operated apart from government control—instead it was run by an appointed Board of Governors. The BBC also adopted the "regional scheme" approach with both local and national programming. This allowed a greater range of reception and also made it possible for people with better quality receivers as well as crystal sets to tune in. In a further attempt to increase listenership, the BBC wanted to broadcast on shortwave outside the country to the farthest reaches of the British Empire. Director General John Reith petitioned for a license for shortwave frequency broadcasting and was awarded it from the post office in 1926.

Radio in the early days was largely a Western phenomenon. In India, for example, wireless telegraphy was slow to appear. The first amateurs started in 1921 and by 1923, there were still only a few dozen sets, private or public, in operation. The first public broadcasting station did not appear until 1935 when the vice president of Mysore University set up a shortwave system with some enthusiastic students and amateurs called VU2HK. The station broadcast entertainments and traditional Indian music. As many governments did, during World War II the amateurs were shut down. After the war, the Indian government began to issue licenses again but few were taken up. It was not until the later part of the 1950s that widespread commercial broadcasting began on the subcontinent.

By the end of the 1920s the BBC, despite the fact that it was then broadcasting around the world, was receiving competition in its own backyard. The BBC had prided itself on the high intellectual content and cultural value of its programming. Some listeners had grown a bit weary of all the high culture and longed for something a bit lower (and more entertaining) on the scale. The core of the network's content, besides news, was classical music, opera, and literary readings, intellectual talk shows on science, politics, and similar material. In 1929 just across the English Channel in France, a station known as Radio Normandie began broadcasting. The next year, the station began to augment its French language content with some limited English language programming. The English language content was provided by a company called the International Broadcasting Company started by a Philco radio salesman named Leonard Plugge. These shows were far more lowbrow and, in some people's estimation, more enjoyable than the more formal and stuffy BBC fare. The station was advertising-supported: many of the commercials were for Philco radios, and some of the first advertisements appeared for automobiles, like Jaguar, as well. Initially, only broadcasting a few hours at night in English, Radio Normandie grew in popularity and expanded its broadcast times to meet the listener demand. Fighting back, the BBC launched its own broadcasts in French, later added Italian and German programming in 1938 on the eve of World War II.

RADIO AND ITS DISK-CONTENTS

Once the technology of radio had been more or less worked out and networks established, attention was turned to what to broadcast. The early radio stations did everything they could to avoid playing recorded music because they did not want to pay a royalty to the musicians who recorded it. As a result, most of the content on the early radio stations was live. In 1923 the American Society of Composers, Authors and Publishers (ASCAP) began to push for mandatory royalties for the playing of recorded music. In an effort to fight back, a group of radio station owners formed the National Association of Broadcasters (NAB), and the two groups began negotiations in order to protect their respective clientele. This in turn led to some of the turning points in radio's growth. In 1925 WMAQ worked out a deal with the Chicago Cubs to broadcast live voice play-by-play coverage of professional baseball. The next year, the Chicago White Sox and football games played by teams from Northwestern University and the University of Chicago were also broadcast.

The 1920s was one of the great periods of social change in the United States. The Roaring Twenties was also the period of the "New Woman," the "New Negro," and the "Harlem Renaissance." Many people, including throngs of new immigrants, began to exercise their right to self-determination, social justice, and all the other advantages of democracy, which had long been denied to them. For these people, the decade was one of excitement and endless possibilities. The Old Guard, the upper middle-class white men of Anglo-Saxon and Nordic background in charge for so long, viewed this time with dread and worked to reverse its effects. One tool they sought to exploit for this purpose was radio. Radio could be an instrument of cultural unification, they thought, which would smooth over the rough edges of the new immigrants and make the New America more in line with the traditions the Old Guard was comfortable with. As such, radio was used to establish racial boundaries as well as economic ones.

The place radio held in the 1920s in the question of race is complex. During these years, the public "face" of radio was white. The owners, announcers, and advertisers were all white, yet a good deal of the content was black or at least black inspired. Roxy, for example, was a white man affecting a Southern, rural black persona. He was working in the tradition of the minstrel show: whites playacting as black by painting their faces. Jazz music was easily the most popular form of music then being played, yet the African American experience, and African Americans themselves were all but invisible. The younger white generation was no longer satisfied with their parents' music or culture and gravitated toward the radical new sound of jazz. But most were just playacting like the minstrels. While some whites did embrace African Americans as people, most wanted only the fantasy not the reality. As jazz evolved into Big Band music, the same held. The story would repeat itself decades later with the arrival of blues and then rock 'n' roll when it took over the radio. Black jazz and then rock in its purer forms were popular but relegated to "race records" and "race stations." When the white variant of Big Band appeared, it exploded on the radio as a more acceptable form. In this way, the creators of a cultural phenomenon were separated from their creation and disassociated from it completely. When black musicians or performers did appear on "white" radio, they were often presented as something unusual and exotic: as if they were Martians completely outside the normal run of "real" America. Minstrel show–inspired radio programs were common and, like the music, were played by white actors. When black performers portrayed black characters, they were expected to "black up," to overemphasize and accentuate the stereotype of voice and behavior.

The epitome of this phenomenon was *Amos 'n' Andy*. The precursor to that show was the *Sam 'n' Henry Show* on WGN, which began airing in early 1926. It came on every night at 10 PM and ran for 10 minutes. The two white actors, Charles Correll and Freeman Gosden, spoke in Southern black dialect. The *Chicago Tribune* decided that a minstrel-inspired show might draw listeners and so hired the actors for a show to be called *The Gumps* (based on a comic strip). Correll and Gosden did not like the Gump concept and suggested something a little different. The *Tribune* liked their suggestion and *Sam 'n' Henry* was born. The newspaper capitalized on its popularity with merchandising and advertising. The rival *Chicago Daily News* station, WMAQ, too wanted a popular program and so hired away the two actors. They could not use the name and so they came up with the names Amos and Andy. The show first appeared on March 19, 1928. It opened with the two heroes arriving in Chicago for the first time and searching for a place to live in the heart of the African American community of South State Street. In order to join in with the life of the community, the two men join a fraternal order called the Mystic Knights of the Sea, whose leader, Kingfish, became as popular a character as the two stars. They start a taxicab company and the stage is set. The show was the first to achieve a kind of rough syndication and found a sponsor in the Pepsodent toothpaste company. Unlike other corporate sponsors who required their names to be part of the show's title, Pepsodent took a low-key role using only short "spot" breaks to advertise its product. With that the radio "commercial" was born. In August of 1929 the show was picked up by NBC.

Amos 'n' Andy both consciously and subconsciously helped establish what it meant, in the eyes of wider white America, to be black. Radio forced performers to exaggerate their acting in order to establish a character as a particular ethnic group. As you could not see them, performers had to act black. Correll and Gosden affected broken speech patterns, poor grammar, and confused syntax in order to sound black, at least the widely held white stereotype of black speech. References to gambling, poor understanding of the value of money, and a tendency to womanize were all tactics and cues used to identify the "blackness" of the characters to the audience. At the same time *Amos 'n' Andy* was a self-contained world. All the characters were, despite the fact they were portrayed by whites, black from a bum to a cop to a business owner to the local millionaire in a mansion. White characters rarely appeared. In this way, the important questions of race never had to be addressed and the comedy could go on untainted by drama. As such *Amos 'n' Andy* was a fantasy which did not contend with the harsh realities of the African American experience. Ethnicity, except where it lent itself to comedy, was rare on radio during this period. As were genuine black voices missing, so were just about any strong ethnic presence. Some local programming had natural local dialects but the networks consciously scrubbed the sound clean to accentuate a nonethnic white sound. Cultural hegemony meant a national, not regional way of speaking. When ethnic voices appeared in shows like *The Goldbergs* (Jewish) or *I Remember Mama* (Swedish), their ethnicity was sanitized to a point where it was quaint and unthreatening.

Amos 'n' Andy reached an enormous audience and ran for over 30 years. The show was a comic serial in which a cast of colorful (no pun intended) characters revolved around the two main stars. Each show had a basic narrative but without the need to extend that specific story from one episode to another. The humor revolved around word play more than any slapstick physicality. It was a successful format that was popular amongst producers as well as listeners and led to the modern "soap opera" and situation comedy (shows from *Days of Our Lives* to *Seinfeld* can trace their origins here). As a historian of radio, Michele Hilmes said, "*Amos-n-Andy* proved to the emergent radio industry that serialized narrative drama, whether comedy, adventure, or romance, could be the basic building block of a new medium" (1997, 82). In the wake of *Amos 'n' Andy*, a host of similar shows followed. Newspaper comic strips proved a particularly lucrative source for material, including *Gasoline Alley*, *Little Orphan Annie*, and *Dick Tracy* (which included a futuristic wristwatch radio).

CONCLUSION

By the 1930s radio had become an integral part of American and Western European culture and was well on its way to doing the same with the rest of the world. Its content ran the gamut from news to politics to religion to entertainment and was paralleling the state of society with its own. After a rocky infancy, radio experienced the rapid growth of adolescence. It had gone from a curious and ill understood form of communication, thought only useful for the shipping industry to worldwide phenomena that had become a central element in human life. From those early private days, radio had taken on the form of cultural icon. No longer at the outer edge of society, radio had become its public electronic heart. It was now time to see what radio could really do with its life.

5

The Cultural Juggernaut

The 1930s saw radio grow as a political animal. Advertisers and later evangelists saw the power of radio to reach adherents and generate followers. Politicians learned the lesson and used radio to spread ideologies and draw supporters. They would use the radio to spread not only words of community and pleas to not give in to fear and calls for individual and collective strength for the common good, but also words of hate and calls for mass hysteria and violence. Paralleling what was going on in society, radio helped bring out the best and the worst in people. By this time, radio had become a constant companion to millions around the world, particularly in the United States. Twenty-six million American households then had radios that were tuned in for 5–6 hours of listening a day. Radio became a phenomenon of communication unlike any before; it became a linking agent of society. Prior to the twentieth century, most people around the world rarely interacted with anyone outside their immediate community. But later, radio connected them to the four corners of the world. Some philosophers saw these developments as inherently positive. Radio, they said, would usher in a new world in which community would rule with cultures brought together for the greater good. Radio would be the ultimate town meeting. Others, though, saw darker implications for the spread of malignancy. Radio assisted in the birth and spread of popular culture. Music had always been a central content of radio broadcasting, beginning in the 1950s and then through the 1960s on, pop music was what most listeners had on

By the 1930s radio listening had become a part of everyday life. Harcourt Index.

their minds. From rock 'n' roll to rap to hip-hop, music did more to bridge cultural gaps than anything else.

A QUESTION OF CULTURE

As commercial radio grew in popularity, the U.S. government felt obliged to form a body that would oversee it. That body became the Federal Communications Commission (FCC). As an "independent executive agency," the FCC grew out of the Communication Act of 1934 and replaced the original Federal Radio Commission (FRC). As part of President Franklin Delano Roosevelt's New Deal program to ease the burden of the Great Depression, the FCC was to monitor the spread of radio and telegraph communications across the country. Roosevelt (FDR) understood that communication would be a vital part of the nation's economic recovery. The communication of crucial news information as well as entertainment (not to mention the transmission of government edicts) would help bring the country together as a closer community and foster the notion that all Americans were bound together in the struggle. The FCC melded communication rules and regulations from the FRC, the Interstate Commerce Commission, and the Postmaster General's office into a uniform set of guidelines. Responsible for telephone communications as well, the FCC's purview was later extended to television. Its primary responsibility was to assign radio frequencies, regulate station wattage and other mechanical concerns, issue broadcast licenses, and more vaguely protect public communications. The commission had seven members, all appointed by the president and ratified by senate vote, who served for 7-year terms (in the late twentieth century the commissioners were reduced to five serving 5-year terms).

While official regulation was limited, there were unofficial forms of regulation going on as well. The arbiters of taste in America had tried hard to keep radio a "cultured" medium. Their stress on serious theater, classical music, and drama was never completely successful. *Amos 'n' Andy*, *Dick Tracy*, jazz, rock 'n' roll, and later hip-hop and rap kept Mozart, Vivaldi, Wagner, and the great dramatists and authors at bay. The evening became the time for cultured, "highbrow" radio while the daytime was the realm of the "lowbrow." People had a chance to sit and listen intently at night, whereas during the day the radio acted as a background to the drudgery. Radio walked many fine lines; one was trying to mollify the interests of the highbrow sponsors and making a living off lowbrow content, which brought in the listeners in larger numbers. (In the early twenty-first century this discussion would become almost moot as highbrow culture all but gave up on notions of being a leading social entity. Classical music radio stations, while profitable, entered the minority as far as number of stations were concerned, while serious talk and intellectual discussion retreated to public radio, and drama all but went extinct.)

While comedy, in the form of shows like *The Jack Benny Program*, *Burns and Allen*, and others led the ratings, dramatic radio was also growing. These shows centered on readings of "great" works of literature, original dramatic writings, and even spoken word versions of popular movies. Shows like the *Lux Radio Theatre*, *Hollywood Playhouse*, and *The Mercury Theater* were held up as the jewels of radio's crown: prestigious presentations that gave the networks an air of class and sophistication and shows that had gravity and seriousness. They were the high culture of radio. The dichotomy between high and low culture on the radio in the 1930s and 1940s was a continuation of struggles that had been going on since broadcasting's earliest moments and mirrored the struggles going on in American society. The rapidly changing aspect of America during this period pitted established Anglo-Saxon corridors of power and hegemony against the enormous influx of immigrants. The Old Guard sought to lessen the blow of this tidal wave of new people by Americanizing them. Serious culture in the form of classical (European) music and literature, and the values and traditions thought to go with them were seen by some as the epitome of Western civilization. These forces abhorred the raucous new cultures brought over from places like Ireland, Italy, and Russia, as well as African Americans and women getting their first real taste of cultural power. As an elixir to this supposed decline, high culture was set up as a yardstick for new arrivals to judge themselves by and an ideal to aim for in reconstructing their lives more in line with their new home. The world of highbrow culture was thought to be able to scrub the vulgar, dirty immigrants clean. Radio's nighttime (prime-time) fare was homogenized as a way not only to teach new Americans how to speak and behave, but also to avoid offending a national audience of many different stripes. Such offence, it was feared, might lead to a drop in advertising revenue as an insulted audience would refuse to buy the products advertised. Highbrow culture was Anglo-Saxon, anything else was ethnic and racial. Therefore, even though the growing American working and middle class was diversely ethnic, radio steadfastly refused to acknowledge it. Taking the lead of Chicago's WGN, the big networks set themselves up as self-appointed arbiters of culture and bastions of goodness, light, and all that was proper on radio.

Listeners instinctively adapted to the notion of disembodied voices coming out of a box and entering their homes. People related to radio as if it were a living thing. The box itself disappeared and the listener became one with those inside. They treated them like close and trusted friends, and believed what they were told. Listeners could identify with the events and dramas coming out, and the line between the real and the imagined blurred. An early example of genuine human drama on the radio was the crash of the German transatlantic airship *Hindenburg*. The *Hindenburg* was an enormous balloon-like flying machine, a sort of luxury liner of the air. With large swastikas painted on its tail, the *Hindenburg* was a flying billboard for the Nazi regime and a showpiece for the Hitler government. Designed to make regular passenger trips between Europe and the Americas, the sight of so large a machine floating silently over the landscape was impressive. On May 6, 1937, the *Hindenburg* was scheduled to arrive at Lakehurst Naval Air Station in New Jersey. Bad weather delayed the landing and so many of the people who had come to watch had already left. Of the people who remained in the light drizzle were radio announcer Herb Morrison of Chicago's WLS and his engineer Charlie Nehlsen. Morrison was describing the arrival of the great machine when at 7:25 PM, it suddenly exploded into a huge fireball and crashed to the ground killing dozens. In the panic, Morrison kept going. (He wasn't actually live on the air, he was speaking into a recording device.) As the explosion occurred, he shouted, "It's burst into flames . . . Get out of the way, please, oh my, this is terrible . . . Oh, the humanity and all the passengers!" He could clearly see passengers falling and jumping out of the ship, some on fire. After the incident, Morrison's recording was played over the air on WLS and then NBC. The normal hesitancy not to play recorded material was dropped in the face of such a momentous event. Though recorded earlier, the *Hindenburg* crash was an example of how real-life drama could make it to the air and draw audiences who wanted to hear about it.

THE PRIEST AND THE PRESIDENT

The 1920s saw the rise of a social movement known as Christian Fundamentalism. This was made up of people who felt American society had strayed too far from the fundamental forms of Christianity. Mostly, the Christian Fundamentalists opposed whatever they thought diluted proper Christian doctrine. Preachers across America took to the airwaves as a way of spreading the Gospel. While some radio preachers found local success, a few found international fame. The use of technology like radio held an odd place in this story. Radio was a "modern" element, yet it was employed by preachers to spread their ideas about the problems of modern life. One idea that was not viewed with such ambivalence was the teaching of evolution in public schools. For Christian Fundamentalists, evolution was false doctrine because, in part, it called into question a literal reading of the Bible.

Fundamentalism, evolution, and radio, all came crashing together in 1925 in the guise of the Scopes's "Monkey Trial." The state of Tennessee had that year passed the Butler Bill, which outlawed the teaching of evolution in the state's public school system. The fledgling American Civil Liberties Union (ACLU) brought a test case to defeat the law and the town of Dayton, and local high school teacher John Thomas Scopes volunteered to be the object of the trial. Scopes was charged with breaking the Butler Bill

Listening to the radio on Sunday morning, 1923. Religion found a powerful outlet on the radio. Courtesy of the Library of Congress.

and the ACLU lined up legendary attorney Clarence Darrow (1857–1938) to defend him. The prosecution employed three-time presidential candidate and well-known Christian Fundamentalist William Jennings Bryan (1860–1925). The stage was set for a battle of legal Titans fighting over a topic of great interest. The trial was a media event and dominated the press for weeks. The Scopes trial was the first such legal proceeding broadcast live on the radio. Chicago station WGN and the *Chicago Tribune* made sure everyone knew that that was a major milestone. The use of radio to cover the case, they said, was radio being used to perform a public service. While superficially a question of religion, the Scopes trial had little to do with religion, or religious broadcasting: it was more a trial of authority and the right to teach unpopular ideas. Despite this, there was much genuine religious broadcasting going on.

One of the major radio evangelists of the era was Aimee Semple McPherson (1890–1944). Flamboyant and controversial, McPherson was a broadcaster when few women could be found on the radio. Born in Canada, her mother was a Salvation Army worker. Always religious, Aimee experienced something of an intellectual crisis as a teen when she was exposed to Darwinian evolution theory. She sought council from a local preacher Robert Semple, who calmed her doubts and then married her (she was seventeen years old). In 1910, the newlyweds headed for China as missionaries. Not long after, Robert Semple died, leaving Aimee alone with a newborn child. Making her way to New York, McPherson bundled her daughter and her mother into a car and headed West to begin a career as a traveling evangelist. She drew a large following with her unique and dynamic preaching style, which included a good deal of the theatrical. In 1921 her career received an enormous boost when she was rumored to have cured a wheelchair-bound woman. By 1923 her ministry had grown enough financially that she opened an opulent, million-dollar temple in Los Angeles. The next year she began broadcasting her sermons over Los Angeles KFSG radio. She used her flamboyant lifestyle to emulate that of the Hollywood community, which was nearby. She cultivated an upscale, urban audience with the radio and her shows were soon reaching national listeners.

In 1926 scandal hit her radio ministry when in May, while on an outing to the seashore, she went missing and was feared drowned. Her followers were heartbroken and a major search ensued. A month later, she suddenly reappeared in Mexico, claiming to have been kidnapped. It was discovered that she had run off with her radio engineer, Kenneth Ormiston, for a romantic interlude. Both sides denied the accusation. Ormiston went home to his wife in Australia and McPherson went back on the air. Following the disappearance event, her popularity was stronger than ever. The scandal only brought in more listeners. She stayed on the radio throughout the 1930s and World War II years until she passed away in 1944. Her career showed that a unique personality came over the radio well, as did religion. She would be followed by a long line of such broadcasters.

The Great Depression of the 1930s had other than simply economic effects. The world experienced a psychological breakdown as well, and dark forces began to grow. In times of such upheavals and uncertainty, people looked for easy answers to complex problems and scapegoats to blame their troubles on. The radio was ready to supply both. Scapegoating is a phenomenon where people latch onto an easy target to vent their frustrations and see in the target the blame for all their ills. These targets usually have little or nothing to do with the cause of the crisis, but that doesn't matter. The scapegoat becomes the symbol to unleash pent-up hostility and fear upon. In these desperate hours, unscrupulous individuals can take advantage of and capitalize on people's fears and prejudices. One of the first to do so during this period came in the unlikely form of a Catholic priest.

Most radio evangelists were content to spread the word of God, draw in the faithful, and chastise sinners within the realm of traditional church practice. Some began adding political elements to theology. The most controversial and influential of the 1930s radio evangelists used broadcasting so well that he came to be known as the Radio Priest. Charles Coughlin (1891–1979) was born of Irish descent in Canada, where he was trained for the Catholic priesthood. In 1926 he was sent to the Detroit suburb of Royal Oak, Michigan, as parish priest for the Church of the Little Flower. He soon found his parish harassed by local Ku Klux Klan members. The feisty Irish priest fought back by giving a series of radio sermons in which he promoted religious freedom. His style of oratory, a deep bass voice that he consciously modulated for effect, proved a hit, and he continued broadcasting. His radio audience grew as did his church's attendance. His sermons concerned the plight of the working class and farmers. He saw big business as inherently evil and a tool of communism and argued that a cabal of capitalists, communists, and bankers were destroying the country. Initially, Coughlin was a supporter of President Roosevelt's New Deal policies, but as the Depression wore on, he targeted the government's reforms as part of the problem. The rhetoric was successful; by 1930 his broadcasts were being sent all over the country and had a listenership in the 40 million range.

As time went by Father Coughlin's rhetoric took on an increasingly dark tone. Always sure of his own righteousness, Coughlin's sermons became more dictatorial, arguing that he held the key to America's success, that the Roosevelt administration was a minion of Satan, and that the New Deal just a ruse for making America a communist state. One of the hallmarks of Coughlin's sermons was that they were oddly vague. While championing "democracy," he was never quite clear just what he meant by the term. He made accusations against "private hands" and "private banking" being the root of all evil, but like democracy, he never explained what these terms meant. He allowed his listeners to fill in the gaps for themselves in whatever way they wanted. That way, his vagueness worked to advantage. As the 1930s wore on, he became more conspiratorial. His turn on Roosevelt and the New Deal may have stemmed from angry disappointment: he felt his support of them would lead to a place of power in the Roosevelt administration; when it did not, he unleashed his wrath on his former ally. He seems to have been more interested in personal power than ending the Depression. His sermons also had an anti-Semitic edge to them saying that Jews were agents of bankers and communists. He was at the same time a populist willing to use fear tactics to scare his parishioners into obedience. He lashed out at all points of the compass, his sermons turning to screeds. By the end of the decade, he had alienated other Catholics worried by his increasingly angry rants. He abandoned reasoned arguments against the rapaciousness of big business and corporate greed in favor of unfocused emotional calls to bigotry and hatred. With the outbreak of World War II, and after some awkward praise of Hitler, Coughlin's local bishop, for years quiet on the subject, finally stepped in and ordered the Radio Priest to stop discussing political subjects in his sermons. He continued to broadcast for years after, but the controversial aspects were gone as was most of his audience.

Following Marconi's lead, the U.S. government began training radio operators. Here African American men in the New Deal program, Civilian Conservation Corps, learn wireless telegraphy at Kane, Pennsylvania, 1933. Courtesy of the Franklin D. Roosevelt Library Digital Archives.

As a way of alleviating the crushing effects of the Depression, President Roosevelt instituted sweeping reforms that he called the New Deal. By expanding the role of the federal government in assisting the citizenry, Roosevelt sought to fulfill a trust between the government and the people. To spread his message of hope, Roosevelt turned to the radio and used it to get his point across, to rally followers to his cause, and effect social change. That he chose this method was an indication of how radio had grown and become an integral part of society. His goal and approach, however, were different from that of the Radio Priest.

Roosevelt knew that something had to be done to reduce the emotional stress as well as the financial effects of the Depression. To do that, he took to the radio in an unprecedented series of broadcasts known as the Fireside Chats. He went on at a time, usually Sunday nights at 10 PM, so that the greatest number of citizens would be able to listen. He wanted the addresses to have a homey, intimate feel, hence the visual imagery of sitting around the family hearth talking. Roosevelt explained how things were going and what the government was doing to alleviate the crisis. The idea was not to "address" the country, but talk with its citizens. He made the first of these broadcasts on March 12, 1933, and made more than thirty of them during the decade. It was an attempt to shorten the gulf between the office of the president and the average citizen. He rarely placed blame or pointed out

Franklin D. Roosevelt having a Fireside Chat in Washington, D.C., April 28, 1935. Radio brought politics and national concerns into citizens' homes. Courtesy of the Franklin D. Roosevelt Library Digital Archives.

enemies; instead he concentrated on getting through to people and calming their fears. Roosevelt discussed the state of the Union in a stark and honest fashion and used nontechnical language. He tempered the bad news with a calm and intimate-speaking voice. Like a radio host, he used anecdotes and established narratives—like a baseball game—to explain the complex forces involved. He stressed that while the government programs of the New Deal were helping, every citizen needed to pitch in. He argued a kind of "we are all in this together" approach designed to counter the feeling many people had that they were alone and forgotten and told them not to panic. The idea of the Fireside Chats was to have the president literally go into people's homes and assure them that while times were tough, they would make it through. Though most never met the man, many people felt they had come to know the president from these radio broadcasts.

The Fireside Chats were one of the great moments in the use of radio to promote peace, stability, optimism, and community for positive purposes. Those philosophers and commentators who saw radio as a great unifier and spreader of democracy saw this moment as a vindication of utopian ideals brought through technology. Radio, however, like all technology is a neutral device. There is no inherent good or evil in it: that comes from who uses it and how. Franklin Roosevelt used it as a way to bring people together in a struggle for good; other politicians would use it just as effectively, but for the spread of hatred and war. There were different wars on the radio, some were fictional, some all too real.

THE WAR OF THE WORLDS

One of the other great radio events of the 1930s in America was dramatic but fictional. Its impact was heightened by the fact that people had become so close to their radios and what came out of them that many thought, at least initially, that it was a genuine event. Orson Welles's *Mercury Theater* had developed a reputation for doing dramatic and effective versions of great literary works as radio dramas. In 1938 he and his chief writer John Houseman hit upon the idea of recreating British writer H. G. Wells's *War of the Worlds* as a Halloween production. Radio shows in the 1920s and 1930s, even the highbrow ones, tended to be 10–15-minute affairs, occasionally a half-an-hour. With the appearance of dramatic radio companies like *Lux* and *Mercury*, hour-long formats were experimented with. These were not serials but self-contained works. There was often a single creative mind that helmed the project instead of a traditional host. The creators of the hour-long dramas were more like film or theater directors, unseen but in control.

Dramatic radio during the 1930s and 1940s saw the rise of superhero serials that were episodic narratives that contained the same characters and often played out over several episodes. A host of crime fighters proliferated, some were taken from literature, some from comic strips, and some were original to the radio. Traditional characters like Sherlock Holmes were common on both American and British radio. Some, like Sam Spade, were taken from the "pulp" novels of the period. Pulps were paperback novels mass-produced on cheap pulp paper in order to lower costs and raise production. Detective, romance, and adventure stories made up the bulk of the pulp genre: styles tailor-made for the radio.

One of the longest running and influential of the pulp radio heroes was *The Shadow*. Originally part of the *Detective Hour* show that premiered in 1930, *The Shadow* proved popular enough that it was spun off on its own in 1936. The true identity of the character was millionaire playboy Lamont Cranston who had spent time in Asia studying the occult, where he acquired the ability to "cloud men's minds" and become invisible. That handy trick allowed Cranston, in the form of his alter ego The Shadow, to be an effective crime fighter. To make the character more mysterious, it was discovered that Lamont Cranston was actually a daredevil flyer, Kent Allard. Part of the show's success was its signature-opening line spoken with an eerie monotone that turned into a creepy laugh: "Who knows what evil lurks in the hearts of men? The Shadow knows!" Ingenious use of sound effects, like a creaking door, created an atmosphere of mystery and dread. With the growing turmoil in the world, radio serials like *The Shadow* allowed good to triumph over evil, at least in fantasy form.

Early dramatic presentations like *The First Nighter* of 1929 were more than just radio adaptations of Broadway plays (actual Broadway Theater was rarely broadcast live). They attempted to recreate the experience of going there. Each show began with the First Nighter (someone attending the opening performance) walking to the theater complete with street noises, entering the theater, and being shown his seat. It was a simple but effective gimmick to bring the listener into the action. Similar virtual reality shows such as *Fleischmann's Yeast Hour*, with Rudy Vallee as the "host" of a fictional show within the show, were common. The *Lux Radio Theater* was a Hollywood-geared show with the studios sending over stars to promote Lux soap products. The host was the film director Cecil B. DeMille. His presence cemented the relationship between the radio and the cinema, even though DeMille had little or nothing to do with the production of the show other than reading lines like any other actor.

Over at the rival *Mercury Theater*, Orson Welles was already a maverick. He had produced a radio version of "Macbeth" with an all-black cast and a series of critically acclaimed dramatic productions, and served a stint as the voice of *The Shadow*'s Lamont Cranston in 1937. The *Mercury Theater* was in many ways an experiment. CBS gave Welles the go-ahead to produce 9-hour-long radio plays—with complete artistic control—to run at 9 PM on Mondays. CBS heavily promoted the series and its *wunderkind* director and pushed the high culture and literary nature of it. The first show was an adaptation of Bram Stoker's *Dracula*. Unlike many other radio producers, Welles grasped that radio was different from the traditional theater, and so dramatic presentations should not just be sound-only plays. In fact, he called his works stories, not plays. In the next installment of the series, "The Thirty-Nine Steps," Welles did not introduce the story (the usual format), it just began. His narration as host did not come in until almost halfway through the production. Welles was an artist experimenting with a new medium. For the October 30, 1938, broadcast, he chose *War of the Worlds*. Like an awkward child stumbling through the surf, radio was about to discover for itself how to swim.

Welles and Houseman (who later gained fame as an actor) decided that a straightforward reading would not work and so used a newsflash style as if the events were actually happening. In the original, Martians land in England and ravage London; in his version, Welles moved the landing sight to New Jersey so the Martians could attack New York City. The story begins slowly with news reports of meteor crashes breaking into a "regularly scheduled" musical performance. As things grow more ominous the performance is forgotten in favor of live, on-the-spot news reports. What are thought to be meteorites soon turn out to be Martian war machines that wipe out the New Jersey State Police and militia units sent against them and then march on New York with terrible results. The show was so effective that some thought it was an actual attack. Switchboards at the studio lit up and legends were born about farmers attacking windmills and grain silos that in the dark looked like fearsome alien invaders. The public reaction to the show was a turning point in radio history. Some argue that the reaction was a result of anxiety of the growing threat of the Nazis in Europe and the Japanese in Asia. (Some argue that the crazed reactions never actually happened, but were just a product of advertising hype.) The *War of the Worlds* broadcast was so successful that sponsors began lining up to become involved with the *Mercury Theater.* The Campbell's Soup Company landed the position as primary sponsor and so the name was changed to the *Campbell Playhouse.* Then soup commercials had to be worked into the show and the subject matter had to be streamlined to attract an even wider audience. As a result, a creative and original show was gutted and turned into a shill for the movie studios like the *Lux Theater.* Welles soon left and by 1940 it went off the air. While the Martians of *War of the Worlds* were fictional, there were real monsters lurking about and they were going to use the radio as well.

RADIO GOES TO WAR

Technology and war have always gone together. This is in part because technological advance allows humans to kill each other in larger numbers and more efficiently. Radio went to war for all sides in World War II. It allowed for faster transmission of military orders and the logistics required for the kind of massive movement of troops and supplies utilized during the war. Radios were put on everything from airplanes to submarines to tanks to the backs of individual soldiers. All the major leaders of World War II made use of the radio. Roosevelt, Winston Churchill, Benito Mussolini, Joseph Stalin, and Adolf Hitler exploited the force of radio to sway the masses to their side and justify their actions. Hitler's speeches were regularly broadcast and he decreed that any opposing views should be suppressed, so he passed laws forbidding German citizens from listening to non-German broadcasts under pain of prison sentences. He used communications media to manipulate the public to make them believe what he believed and to hate what he hated, and to see an aggressive war of conquest and empire as good for them. Hitler took power in Germany on January 30, 1933, and made his first national radio address the next day. He told his listeners that he was not invading or waging an aggressive war, but that he was liberating countries from oppression and that he was preempting enemies who wanted to destroy Germany. Radio helped him trumpet Nazi party successes, spread hatred for Jews and others, and hide its failures. In 1942, for example, as part of his invasion of Russia, Hitler and the Nazis laid siege to the strategic city of Stalingrad. From their invasion of Poland in 1939 everything the Nazis did had gone their way and they had no reason to believe the Russian campaign would be any different. However, the Russians fought to save their city with great tenacity, and inflicted upon the Germans their first major defeat, turning the battle of Stalingrad into a catastrophe for the German army. To keep people at home from learning of it, a series of faked broadcasts were put together to give a rosy picture of what was going on there. In reality, as the broadcasts were being made, tens of thousands of German troops were meeting their end in the icy wastes of Southern Russia.

Following the October Revolution of 1917, the Bolshevik and later Communist government made quick use of wireless telegraphy to get its message out, not just across Russia but to sympathetic individuals across Europe. Their effort may be the first use of radio communication for political propaganda purposes. Communist leader V. I. Lenin saw radio as an important tool for social control of the Russian masses. As an enormous landmass that technologically speaking was still at a medieval level, Russia was a place where radio could make up for a lot. The problem was that in 1917 true radio was not yet available. A system was set up to send official Communist Party edicts by Morse code to newspapers across the country, which then printed them. Seeing such an important role for radio in promoting the revolution and keeping the proletariat under control, the Soviet government spent heavily on getting voice systems into operation. Despite this effort, there were few people in the country who could afford radios in their homes. The party resorted to the novel idea of putting up extensive public address systems so that people could hear radio without having to have a radio: it also made radio listening a group effort. This allowed for the realization of Lenin's dream that every village should have a radio. As the government controlled the media in Russia, nothing critical of the government was ever heard nor were opposing views. During World War II, Stalin made use of the radio to rally people to defeat the Nazis and to take their attention away from how he himself had ravaged the country.

When World War II broke out in Europe, CBS's all-powerful leader, William Paley, wanted the news department to report on the war and bring the realities of the Nazi war machine home to Americans. He sent a number of reporters overseas to cover the events. Men like Howard K. Smith, Eric Sevareid, and Edward R. Murrow reported back to New York. Murrow (1908–1965), in particular, became famous for his reporting from London while the city was under attack by the Nazis in what came to be known as the Blitz: the Nazi aerial bombing of the city. Hitler's Air Marshal, Hermann Goering, insisted that he could force the British to capitulate by a systematic bombing of the British capital. In a relentless program of air raids, the Nazis hammered the city for weeks, setting large parts of it on fire and generally wreaking havoc. Murrow set up a position on a rooftop and recorded riveting baseball game-like play-by-play of the nightly assaults. The segments would begin with him proclaiming "This is London" and in his calm monotone, Murrow described the scene. Making the reports that much more immediate, listeners could hear sirens, explosions, and the repetitious bump-bump of antiaircraft guns in the background.

Raised a Quaker, Murrow attended Washington State University where he majored in speech. He was hired by CBS in 1935 as an overseas scout rounding up speakers for the radio. Idealistic with a refined sense of integrity, Murrow saw the rise of Fascism, the Nazis, and the outbreak of war in Europe as topics CBS should cover. He felt that the personal touch of a reporter speaking from the heart of the action was more immediate than having an announcer read copy from a studio. Though he actually wrote his reports out beforehand, his speech training allowed him to read them in a spontaneous manner that listeners gravitated to. Not a trained writer, Murrow had an innate talent for turning a phrase and creating an image in the listener's mind. Looking out over the city in the calm before a Nazi raid in 1940, he described "the fantastic forest of London's chimney pots." Murrow's style along with Paley's vision placed an indelible stamp upon radio reporting and became the standard for television as well. After the war, Paley and Murrow, who had become close, drifted apart. Paley began to give in to government pressure more and more. Murrow saw that kind of pressure as undemocratic and resented Paley for not standing up to it the way Murrow thought he should have. Murrow had seen the ravages of the war and the Nazi death camps firsthand and felt resisting government control and demagoguery a way of keeping it from happening again. He eventually left CBS over this issue and went to work for the Kennedy administration. War reporting on radio stirred emotion; music could do the same.

THE ROCK 'N' ROLL REVOLUTION

Even without the problem of royalty payments, record companies were reluctant at first to have their music played on the radio. Concert promoters and vaudeville agents discouraged their acts from appearing on the radio fearing it would hurt record and sheet music sales. Eventually, music producers came to realize that far from hurting sales, radio airplay boosted it. Radio and recorded music entered into a happy, though sometimes rocky, marriage. The main ingredient in this coupling was rock 'n' roll. A hybrid creation of jump blues, jazz, and country-western music, rock 'n' roll came originally out of the African American experience and was picked up on by white musicians in the 1950s. First known as "race" music, the term rock 'n' roll may have come from the 1922 Trixie Smith song, "My Baby Rocks Me with a Steady Roll," and was a veiled reference to sexual intercourse. This music was not played on mainstream radio stations, but became popular amongst young people, both black and white, interested in finding their own voice apart from that of their parents. They were drawn to the rhythmic danceability and suggestive nature of it: it was primal not formal. Rock 'n' roll began to appear on radio when it was copied, or covered, by white artists like Pat Boone, who were deemed appropriate for the medium. These versions had all the sexuality drained from them. As white performers who loved the style began to cover black rock and rhythm and blues (R&B) songs with the same high level of energy as the originals, rock's popularity spread.

With mainstream radio resisting playing original black music, a few white radio announcers began playing it on their own, often clandestinely. The most influential of these pioneering operators was Alan Freed (1921–1965). Beginning his career in radio as a sportscaster in 1942 on WKST in Pennsylvania, Freed passed through several stations. In 1949 he reached a turning point. Working on a Cleveland, Ohio, television station, he was approached by Leo Mintz to host a radio show on WJW, which would allow him to play what he wanted. Mintz had opened a record store in Cleveland in 1939, which specialized in "race" music and other non-mainstream precursor forms like jump blues and jazz and thought Freed's kinetic style would make a good fit to the music. As a radio host, Freed called himself "Moondog," spun hot records, and became very popular. He started using the term "rock 'n' roll" to describe the music and it caught on. His playlists were partial to acts like Frankie Lymon and the Teenagers, electric guitar pioneer Chuck Berry, and white bands like Bill Haley and the Comets (whose "Rock Around the Clock" is the most well-known rock 'n' roll song ever recorded). Freed went on to become the first major rock music promoter and concert organizer.

In the wake of Freed's success, the deejay (short for disk jockey, one who played records) became a fixture on radio. This position came relatively late in radio's history as prior to this period there were no disks for the jockey to play. The first well-known deejay was Martin Block (d. 1967). There had been a few broadcasters who occasionally played recorded music and Block worked for one, Al Jarvis, at KFWB in Los Angeles. Jarvis hosted a show that he called the *Make Believe Ballroom* in the early 1930s where he pretended as if the recorded music was live. In 1934 Block went to New York where he acquired a job at WNEW. He began doing what Jarvis had done, even using the name *Make Believe Ballroom*. The show was a hit and WNEW turned it into a centerpiece of its programming. This helped make playing recorded music on the radio acceptable. By the 1950s the deejay position came into its own. In addition to Freed, others like Dick Clark, Bruce "Cousin Brucie" Morrow, and Murray the "K" set the standard for the radio music personality and were often more successful than the musical artists they played. Brooklyn born Bruce Morrow loved pop music and after graduating from New York University, entered the radio business. While working at WINS, New York, he began to use the nickname Cousin Brucie. He addressed his listeners as cousins as well, bringing them together as a community. His fame grew when as a deejay for WABC, New York, he hosted a series of live rock concerts from the nearby New Jersey amusement center of Palisades Park and when in 1965 he introduced the Beatles at their pivotal Shea Stadium show.

Another well-known New York deejay at the time was Murray Kaufman, better known as Murray the "K." Kaufman (1922–1982) began as a child actor (his mother was a vaudeville singer) and then as an adult went into advertising and radio producing, eventually becoming an on-air announcer. He gravitated to pop and rock music and, as almost a prerequisite for pop music deejays, began building a personality. As Murray the "K" he played original black blues, rock 'n' roll, and Latin music instead of the homogenized versions. Deejays had to do more than just spin records: they had to be part emcee and part performer. He would do crazy stunts on the street or on subway cars, and introduce bands at live events. He became acquainted with the Beatles in England before they came to the United States. When they finally arrived in New York, Kaufman was at their side squiring them around, advising them, and broadcasting from their hotel room. He became so associated with the British phenomenon that he came to be known as the "Fifth Beatle."

One of the technical aspects that helped rock 'n' roll prosper was that in 1949, RCA/Victor introduced a diminutive disk that would play at 45 revolutions per minute (rpm) instead of the standard and larger 78 rpm or 33⅓. The "45" as it came to be known, contained one song per side and was cheaper to make and more durable. As a result, records and record players were more transportable than before as were transistor radios. The recording quality of the disk—at the low end of the scale—was perfect for the simple and often crude rock songs, and for radio play. This was both a boon and a problem for rock 'n' roll music. The popularity of the form brought out the unmistakable sexual and suggestive content of the songs. In parts of the country still suffering the throes of blatant racism, white kids listening to black music was considered offensive. A series of boycotts ensued where morality groups threatened radio stations and advertisers linked to this new music. One place a listener could hear a black performer was *King Biscuit Time*. Beginning in 1941 on KFFA in Helena, Arkansas, the show had live blues musicians, including Sonny Boy Williamson, at 12:45 PM everyday for 15 minutes. Named after the sponsor, King Biscuit Flour, the show opened the way for other black-oriented radio stations and included the first successful black deejay, Early Wright (a form of the show has continued to air into the twenty-first century as the King Biscuit Flour Hour). A number of white radio stations responded to the pressure by voluntarily removing rock music from their playlists and even publicly smashing records—Elvis Presley and the Beatles in particular—to avoid the wrath of the anti-rock movement. In Boston in 1956, city fathers, alarmed at what their children were listening to, put together a special board to prevent any offending music being played on local radio stations. The next year singer/guitarist Screamin' Jay Hawkins's hit "I Put a Spell on You!" sold an enormous number of records, but was banned from the radio because of its erotic nature.

The poster child for worry about pop music, the radio, and obscenity was the song "Louie, Louie." Originally written and recorded by Richard Berry in 1955, the song kicked around for years being rerecorded several times with little notice. In 1963 the song was recorded yet again by an obscure Portland, Oregon, band called The Kingsmen with a lead vocal by Jack Ely. Their version was a departure from the original, turning it from a calypso ballad into an anarchic rock 'n' roll stomp. Unhappy with the crude quality of the recording, the band did not want the song released: it was released anyway and became a huge hit. The song's success, or infamy, stemmed from the fact that Ely's lyrics were unintelligible due to the poor quality of the recording. This allowed listeners to insert whatever they thought the lyrics were. Rumors spread that the lyrics were obscene and explicitly sexual. This only made the song more popular. The state of Indiana banned the song from radio airplay and the Federal Bureau of Investigation (FBI) launched a 2-year investigation that failed to show there was anything objectionable in the song. Ely always claimed that he sang the original Berry lyrics—which are known and which are not sexually explicit (a simple love song about a sailor and his faraway love). Rock music brought the question of censorship of the radio out for the first time: and would never really go away. Along with suggestive lyrics, rock 'n' roll had another problem. In the late 1950s and early 1960s the word "payola" came into usage.

Payola refers to the practice of record companies paying deejays to play their music on the radio. At first this practice was not actually against the law; bribery was, however. Payola might not have become a scandal if not for the growing rivalry between American Society of Composers, Authors and Publishers (ASCAP) and BMI. In 1940, the federal courts ruled that radio stations could play recorded music without first asking permission of the artists or record companies. In 1942 the American Federation of Musicians (AFM) launched a boycott of radio stations that were not paying royalties. Record companies wanted their records played, so they quietly started paying for it. As a result, a bidding war broke out between ASCAP and BMI. ASCAP was formed in 1914 to protect the rights of songwriters, authors, and publishers by registering their work and helping with copyright and royalty fees. An early lawsuit brought by founder Victor Herbert was decided by the U.S. Supreme Court, which ruled that if a business was to use someone else's music for the generation of profit, the musician had to be compensated. ASCAP became the home to mainstream-established artists and writers. While the organization did an important job to protect artists' rights, it became somewhat restricted because of the special requirements for membership. As a result, many musicians were left out. To meet that need, BMI was formed in 1939. ASCAP looked after established forms such as classical, big band, and vaudeville, as well as print authors. They represented the conservative, established music makers back to the 1920s. BMI became a magnet for not only country-western artists, but also the socially subversive rock 'n' rollers and black performers. The new generation quickly came to dominate the airwaves, thanks to deejays like Alan Freed and others, and rock and related pop music pushed the older forms off the charts and off the radio, and ASCAP resented it. Many of the old-line performers were openly hostile to the new music. As BMI's influence grew, it came to control a significant percentage of the pop music world: the same music which had come to dominate radio play by 1960. Following the TV quiz show scandal of the 1950s when popular game shows were found to be rigged, ASCAP prodded the government in 1960 to investigate

Alan Freed began to use the phrase "rock and roll" in the 1950s and helped inaugurate a cultural revolution spread by the radio. December 1956. Copyright Bettmann/CORBIS.

the payola practices of BMI, claiming it was bribery. ASCAP artists hoped that the payola investigation would sound the death knell of rock 'n' roll and Elvis Presley. The Federal Trade Commission began to look at radio stations and record companies, which, fearful of the heat being turned up, turned on the deejays. They gave up the names of any deejay who had accepted their money to the government. Several key disk jockeys were singled out, including Murray the K, Dick Clark, and Alan Freed. Clark appeared before the investigating body producing a series of charts and statistics to explain how, while he did have a financial stake in many records, that he was not taking bribes. His ploy worked and he was let off with little more than a warning. Clark came off as a clean-cut, respectable young man, not a wild radical, so his career was unaffected; and he went on to be a music promoter and creator of the long-running television show *American Bandstand*.

The great loser in the payola scandal was Alan Freed. Unlike Clark who cooperated with the government, Freed resisted and admitted taking money. He had made many enemies in the record business and in political circles by insisting on playing original black and rock artists instead of the homogenized corporate versions, so he had little support and fewer friends. He argued that his payments were a form of consulting fees from record companies. To protect themselves, his employers at WINS fired him. His case came up in 1962 and he pled guilty to bribery and paid a small fine, but his career and reputation were ruined. In 1964 he was indicted for tax evasion. A broken, unemployed, and abandoned man, Freed died in 1965 from liver problems. Congress continued to investigate payola and radio into the mid-1970s even resurfacing briefly in the mid-1980s. Payola was ultimately about which records were played on the radio and when. Payments were made because there were far more records available than there was airtime: and airtime meant sales. In response, a Nebraska deejay and radio producer, Todd Storz (1924–1964), invented the Top 40. Instead of allowing the deejay to pick songs, musical content would be picked for the deejay based upon sales and other intangible elements. This way the deejay could not, theoretically, be bribed to play certain songs. Top 40 came to dominate radio airplay.

Rock 'n' roll songs were particularly well suited to Amplitude Modulation (AM) radio play and the Top 40 concept. Whereas classical music and big band were complex forms of music, rock was musically simple. Instead of the large horn sections of a big band, rock bands might have one saxophone. Classical orchestras had many instruments producing subtle and nuanced sounds, while rock bands had a ragged electric guitar leading the way. This simple and raw sound recorded well on the crude audio equipment of the day and lost little in the way of sound quality on the equally crude AM transistor radio receivers. By the middle of the 1960s, rock music began evolving musically beyond its humble origins. The Beatles in particular began experimenting with strings, classical horns, feedback, and other unusual studio manipulations as well as stereo recording. Impressed by the results, other bands began doing the same. As the 1960s turned into the 1970s, rock 'n' roll was becoming rock, a new and more "artistic" form. The more complex recordings needed a medium that was better suited to transmit the sound faithfully. Enter Frequency Modulation (FM).

FM RETURNS

From its inception, FM radio had remained in the background. In the late 1960s and 1970s, it finally came into its own. FM deejays eschewed the kinetic energy of their AM siblings and went for a more self-consciously artistic demeanor. They played folk music and traditional blues, and then the more complex rock of the "British Invasion" bands, "Glam Rock," and all the other material that would never have been played on Top 40 AM stations. This was partly due to artistic choice, and partly due to the fact that the government said FM stations could not simply copy the content and format of AM. One of the AM deejays to jump to FM was Murray the K. On WOR-FM, Kaufman began playing not just album cuts, but entire album sides. Album Oriented Rock (AOR) was a rejection of the Top 40 format. Long-playing (LP) records were ignored by AM radio. The FM "jocks" knew that the new experimentation taking hold of music was producing interesting and exciting sounds that were on albums, not singles. From this was born the "Free-Form" approach. Top 40 only played the hit 45 singles and ignored albums. Not under the constraints of Top 40, FM stations began to play other than simply "hit" music. This meant longer songs could be played. The typical Top 40 song ran from 2–3 minutes. Conventional wisdom said listeners would not stand for anything longer. FM rock stations played songs like Led Zeppelin's "Stairway to Heaven" and Queen's "Bohemian Rhapsody" that ran far longer. A major moment in lengthening pop music for the radio was Iron Butterfly's "In-A-Gadda-Da-Vida" at an astounding 17 minutes in length. These songs not only broke the mold of traditional radio fare, but also became enormous hits. The free-form honeymoon of FM did not last particularly long, however. As FM became increasingly popular and revenue-generating, it became more corporate, formal, and bland—as had rock music in general. In the 1980s, the dominance of corporate FM radio was challenged when the original free-form, noncommercial music movement resurfaced as alternative rock and was nurtured by college and university radio stations, which were less tied to market shares. Budding college deejays played the music they wanted, and they did not like mainstream pop. As a result, bands like REM and the "grunge" bands like Nirvana, Pearl Jam, and others—who would never have had their music played on Top 40—had their careers take off.

FM radio also did something rarely done on AM; it aired live rock music. Broadcasting parts of or even entire performances of popular bands became a staple of FM rock radio. Radio had already proved itself capable of turning obscure musicians with at best local followings into national stars, Elvis and the Beatles being examples. Live radio could do the same.

Growing up in New Jersey, Bruce Springsteen aspired to rock 'n' roll stardom and began playing with a circle of musicians in the small seaside town of Asbury Park; a Gilded Age resort destination of choice, which by the 1960s had fallen on hard times. After a few hardscrabble years, Springsteen met music industry legend John Hammond (1910–1987), the man who helped bring swing music to the radio and had discovered an astonishing array of artists, including Aretha Franklin, Count Basie, and Bob Dylan. Hammond, impressed by Springsteen's lyric abilities and gritty unpretentiousness, signed him in 1973 to Columbia Records. The result was *Greetings from Asbury Park, New Jersey* and a follow-up effort, *The Wild, the Innocent, and the E Street Shuffle* (1974). Both works received critical acclaim, but did not perform well financially. Columbia executives feared Springsteen was destined to remain little more than a local favorite and reluctantly gave him a last chance to redeem his career with a third set of songs. Springsteen and his E Street Band had developed a loyal following based upon his working class-related lyrics and the strength of the band's live shows. Part performance art, part story hour, part revival meeting, and all rock 'n' roll, E Street shows went on for hours, yet were invigorating in the way music was supposed to be. Critics held up Springsteen as an elixir to what they saw as the superficiality and blandness of corporate rock and the soullessness of the disco era. On stage, he would extend songs with spoken monologues about his life or even play one song inside of another. He would often change a song's arrangement from one show to the next. This approach was a major departure from the standard pop shows of the 1970s, where bands played songs from their latest album exactly as they appeared on the album, so that radio listeners would recognize them and buy them, rarely saying anything to the audience beyond "Good evening!" or "Goodnight!"

In a last bid for success, Springsteen and the band entered the studio to record their third album. In the middle of the recording sessions, he did a series of live shows at Philadelphia's Main Point club to work out the material. One of the shows was broadcast live on the radio to the Philadelphia listening area on station WMMR-FM. Included in the set was a song called "Wings for Wheels" that was transformed into "Thunder Road" on the final album and became one of his most popular songs. By the end of the summer of 1975, the new album was ready under the title *Born to Run.* To begin the tour and announce the release of the record, Springsteen and company did a series of shows that August from New York's prestigious Bottom Line club. A popular spot for bands and music executives alike, the Bottom Line was a small club with fine acoustics, which made it a showcase. The show on the 15th of August was broadcast live over WNEW-FM, then New York's premier rock station. The Bottom Line show reached a large and influential music audience: business people, writers, and critics as well as fans. Understanding the make-or-break nature of the performance, the band played with great energy and enthusiasm and the broadcast was a major success. When *Born to Run* was released shortly after, it was immediately acknowledged as one of the great rock 'n' roll albums in the style's history. Many saw live radio broadcasting as the thing that helped save Springsteen's career and launch him into national and then international status.

The other area of explosive radio growth was in the realm of music produced by African American and Latin artists. In the 1970s black music hit the airwaves in a big way for the first time. There were a number of rock stations that regularly played black blues, soul, and MoTown hits, but just a few stations were geared directly to the black and Latin markets. Beginning in the 1970s more stations appeared that played black music exclusively and by the 1980s, such stations were common. Rap and hip-hop began as underground musical movements and then took over the airwaves in the way rock 'n' roll had done 20 years before. As a musical force, rap and hip-hop became increasingly aligned with political agendas. With the rise in rap and hip-hop, questions of race and lyric content came forward again. In 1990 the group 2 Live Crew had its album *As Nasty as They Wanna Be* deemed obscene by a Florida judge. The album was pulled from record shops and radio stations as a result. Rap's offshoot, "Gangsta Rap," pushed the limits of sexual content and violence in ways which made Screamin' Jay Hawkins sound quaint by comparison. Mainstream radio at first resisted playing this music, so a subculture of radio appeared, which played it exclusively. Because of its antiestablishment stance, rap and hip-hop drew an audience large enough to eventually turn it mainstream.

THE BIRTH OF "THE MEDIA"

There have always been people with plenty to say about the radio and what it does. As radio grew as an integral part of society, people began to analyze what was coming out of it. Prior to the 1960s radio was just radio. In 1964 radio, journalism, and television went from being media to being "The Media." This occurred when Professor of English and communications Marshall McLuhan (1911–1980) published the book *Understanding Media*. This groundbreaking work helped establish the discipline of media theory and launched a new way of looking at the world we live in.

McLuhan was born in Canada, attended the University of Manitoba and graduate school at Cambridge, England. As a new professor of English literature, he began to notice that different generations of people viewed the same things in different ways from each other and had different attitudes about the world. The result of this observation was that he began to study various forms of communication and the way they were perceived by people. He attempted a complete reinterpretation of how humans communicated with each other and how mechanical forms of communication—print, radio, and television—altered our understanding of things. In *Understanding Media*, McLuhan invented the philosophical study of what he then termed The Media. He argued that there were two types of Media: cold and hot. Cold Media were those that required a lower intensity of interaction between the media and the user. These included the telephone and television. Hot Media were those that required a more intense interaction and included print, film, and radio. He also argued that to understand human society and how it came to be formed and how it was perceived by people, one had to study the Media itself. He wanted to understand how communication technology interacted with culture and how this forced people to interact with each other based upon how they were told to act by what they saw, heard, or read. He said that the world had become dominated by communications. Radio and other forms were no longer just part of society or just another technology, they were the driving force. This had helped reduce the world to a "global village" (another of his terms). Print had brought this close, but electronic communications had done so to a greater level. He created the phrase "the Media is the message" to explain (as he felt the Media was a shorthand for communications, not a deeper form, it made sense that a phrase could be used as a shorthand to explain a much larger, more complex phenomenon) how a technology like radio no longer simply commented on events, it was the event. What came out of the radio box was less important than the box itself. McLuhan made the forms of communication transmission a topic of research, investigation, and philosophical consideration. That you are now reading a book on the history of radio is evidence of this.

One thing McLuhan focused on was the power of Media advertising to tell people not just what to buy, but what to think about themselves and others, so they would buy specific goods and behave in specific ways. It could construct artificial ways in which people saw things, for example, men and women. Advertising said that in order for a woman to attract a man, she must look and behave in certain ways. In order to do that, they had to purchase certain products. Legs, for example, became a "display object," not just attractive legs, but legs attractive in a specific way. Not the legs of an individual woman, but anonymous mass legs of an artificial ideal. Men then had to live up to their own artificial ideal in order to attract those legs. People were made into faceless automatons, not individuals. And if you didn't conform to the ideal, there must be something wrong with you. McLuhan argued that radio and television and earlier forms of mass communication through time had created images to live up to. The Media did not become like us, we had to become like the Media.

Another part of McLuhan's thesis was that people both extend themselves and amputate themselves with the Media. Radio extends our hearing by allowing us to hear things we would not be able to with our ears alone. The downside of this extension is that there is a corresponding amputation. Radio cut into people meeting face-to-face, it reduced human personal interaction as a result of spending more time with the radio. While radio helped create a global community, it also helped separate individuals. What we do, McLuhan argued, was to emphasize the extensions and downplay the amputations. When FDR came on the radio for a Fireside Chat or when *Amos 'n' Andy* or the *Jack Benny Show* came on, a house full of people came to a standstill and focused on the radio instead of each other. A student of McLuhan's and philosopher of communications, Paul Levinson pointed out that hearing was the key to the birth of democracy in ancient Greece. Citizens were expected to listen to political discourse and comment on it. Greek politics was "defined by the extent to which an audience could hear a speaker's voice-rendering democracy . . . an acoustacracy" (Levinson 1997, 78). As Levinson points out, our cognitive abilities are hampered if we cannot communicate those ideas to others. Ideas and philosophies were already present when radio was born. The radio simply allowed those ideas to filter out to listeners much faster than any previous communication technology.

Media theorists argue over whether there is an inevitable effect of communication technology on human society. That it does is known as "hard" media determinism. The idea is that an information technology once developed will have a specific impact, which can be foreseen before it occurs or that whatever impact it has was meant to be. The flip side is "soft" determinism, which argues that media like radio, despite its impact on our society, are part of a much more complex system involving other elements and that whatever happens is not unavoidable. Radio does not cause things to happen, it only plays a role (along with many other factors) in allowing it to happen. Radio never alters the course of human history on its own, it is how radio is used which does that. McLuhan called for a careful and deep analysis of what media was doing for and to society. He almost single-handedly began the field of Media Studies and to a considerable degree, all the studies of radio, television, print, and film—any type of human artificial communication—is a continuation of what McLuhan started.

CONCLUSION

In the 1920s radio was little more than an infant crawling about. Through the next 30 years, the infant learned to get up and run. The full potential of radio manifested itself during that period. It went from a crude device for ship to shore communications to a mass media capable of reaching and transforming millions as friend, confidant, and master. By the later part of the twentieth century, thinkers and philosophers were beginning to look at radio itself to see what they could find. The image seen was not always a pleasant one. The seriousness of the Depression and war years was supplanted by a preoccupation with pop culture. The end of the century, however, then saw a return to issues and questions other than "Who put the bomp in the bomp-she-bomp?" to questions of society's very future.

6

Haranguers, Listeners, and Howard Stern

By the later part of the twentieth century, radio was reaching maturity: it had gone well beyond the bounds of a curious communication device employed by the military or a few amateur enthusiasts. It had also gone beyond the bounds of a simple entertainment medium. The 1930s and 1940s showed that radio could be a powerful force in people's lives and in world events. It became, for millions, the primary source of information about the outside world. What this meant was that the field was open for other, more philosophical innovations. However, by the 1950s radio had to compete with television and by the 1990s with the Internet. Through it all radio reinvented itself and adapted to changing times in order to stay relevant. The mature status of radio was evident in its move away from sports and music as a primary content and on to such weighty topics as politics, morality, obscenity, freedom of speech, and the very nature of democracy: subjects radio had never addressed in quite this way before.

TALK RADIO

Early commercial talk radio was a one-way communication: a host talked to the audience. (Talk radio is defined here as a show where there is no music or traditional theatrical performance, only discussions of various topics.)

Due to technical limitations, listeners could not engage directly in the discussion. Some stations began to hold in-studio panel discussions in the 1930s and on occasion hosts would read transcriptions of listeners' calls over the air. After World War II, the situation changed as a result of the improvement of broadcast and telephone technology. In 1945 a nighttime musical show host, Barry Gray, may have held the first on-air conversations with callers. If he did, it is unlikely the station—WMCA-AM in New York—approved it. Gray engaged in an on-air phone call from popular bandleader Woody Herman. That one call proved so popular that WMCA management overlooked the breach of etiquette Gray may have made and embraced the concept. Talk radio in its modern form was born.

The changeover from music to talk radio on amplitude modulation (AM) stations, in particular in the 1960s, was due to changes in radio technology and the music industry. Progressive Pop music shifted *en masse* to frequency modulation (FM)'s better quality broadcasting. This exodus left AM with little more than sports and weather. With plenty of space to fill, AM stations turned to talk. The first all-talk stations like KABC in Los Angeles and KMOX in St. Louis appeared in the early 1960s. Despite help from talk radio, AM continued to decline dramatically and may have faded completely if not for the surge in the 1980s of talk "personalities" whose content and concerns veered hard to the right of the political spectrum.

THE CONSERVATIVES TAKE OVER

One of the obvious areas opened to talk radio was for sports commentary and discussion. A godsend for fans of all types of athletic competition, sports talk radio became a popular outlet for those legions who had very strong opinions about why one team was better than the other and how teams ought to have been run. While the topics of talk radio ranged far and wide from sports to cooking to stamp collecting to pets to religion, the greatest impact was in the realm of political talk. Some scholars see the rise of conservative talk radio in America as a result of gains made by the Democratic Party and liberal political philosophy in the wake of the election of Bill Clinton as president. Those who felt trepidation over civil rights, gay rights, women's rights, environmentalism, and a host of other "liberal" causes found voice for their frustration on the radio. Though a number of radio hosts capitalized on this situation, the most successful of the early politically Right leaning, conservative talkers was Bob Grant. Originally from Illinois, Grant began his radio career in the late 1940s after graduating from college. By 1962 he was being mentored by an early purveyor of controversial talk radio, Joe Pyne, who made his mark on radio history by discussing issues from a conservative point of view during the counterculture era and verbally sparring with callers. Taking over for Pyne in 1964, when the latter went on to television, Grant found his niche as a conservative pundit. Eventually arriving in New York in 1970 Grant worked his way through several stations until taking a position at WABC in 1984. It was from this position that Grant gained his greatest notoriety. In the huge radio market of New York City he swaggered with conservative, right wing bravado taking on and denouncing all who opposed him. He took great glee in hanging up on callers, chastising nonwhites for being unfit to appreciate democracy and calling them "beasts" and "screaming savages." The more controversial his ranting, the higher his ratings seemed to go. He was accused of being racist, sexist, and, homophobic amongst other things.

In the mid-1990s Grant became a more involved political activist. He had a parade of New York and New Jersey Republicans through his studio including U.S. Senator Al D'Amato, Mayor Rudy Giuliani, and Governors George Pataki and Christine Todd Whitman. Grant boasted that it was his public support and call to his listeners to vote that put them all in office. Grant's outspokenness came back to haunt him, if only for a few moments. In April of 1996 a delegation from the Clinton administration was in the volatile Balkan region on a fact-finding mission concerning the fighting then going on there and U.S. involvement. The plane carrying the delegation crashed in the mountains of Croatia killing all aboard including the U.S. Secretary of Commerce, Ron Brown (an African American and Clinton confidant Grant had disparaged in the past). Upon hearing the tragic news, Grant quipped that Brown was probably the only survivor "because, at heart, I'm a pessimist." His sarcasm over a deadly plane crash caused a storm of protest and WABC fired Grant. Competing New York station WOR promptly hired Grant and he was back on the air as defiant as ever. Following his firing, Grant began complaining that his right of free speech had been assaulted. Following the 9/11 attacks, Grant was more vituperative than ever, lashing out against Moslems and wholeheartedly supporting the Bush administration's actions in Afghanistan and Iraq. In behavior typical of a growing fanaticism, Grant attacked not only the opposition, but also those ostensibly on the same side of the issue as he, but who were not showing the proper enthusiasm and partisanship. In April of 2004, Grant lambasted the governments of Portugal and Spain, calling them "cowards," for withdrawing their troops from Iraq. Close to home in the same month, Grant referred to Republican former governor of New Jersey and president of Drew University Tom Kean as a "pigeon livered pip-squeak." Kean had been appointed the chair of the special commission looking into the days leading up to the September 11 attacks. The Bush administration had stonewalled the commission's investigation by giving up information only grudgingly, underfunding it, and initially disallowing Security Advisor Condoleeza Rice from testifying before it. Kean and the 9/11 Commission grew increasingly perturbed with how they were being treated and more importantly by what they were finding. Kean was mildly critical of the administration and was, as a result, pounced upon by the conservative radio talkers. Grant declared Kean, and Grant's former ally Christine Todd Whitman, of barely being members of the Republican Party and to having been co-opted by the Democrats.

In the wake of Bob Grant's success as a conservative talk show host, and the backlash against a brief liberal ascendancy in politics, came a legion of conservative talkers. Bill O'Reilly, Laura Schlessinger, and many lesser imitators ranted and railed at the state of the nation, the "Liberal Media," welfare mothers, homosexuals, immigrants, and all the other bogeymen, which haunted conservative nightmares. The most successful of the new generation was Rush Limbaugh. Bringing conservative political talk to an entirely new level, Limbaugh roared with great bombast and drew in an enormous audience as his show was syndicated from coast to coast. He pushed all the hot buttons of American politics in a deep bass voice that spoke to millions of listeners who had felt elated under the Reagan administration in the 1980s, but then abandoned by a country they saw as drifting dangerously to the Left under Clinton. To his credit Limbaugh never claimed to be fair, balanced, or neutral. He argued that "The Media" was so liberal biased that the conservative viewpoint had been lost. (Most of the conservative talkers paradoxically claimed to not be part of The Media.) He was redressing the situation by taking an unabashedly Right stance. He enjoyed pointing out the personal failings and foibles of liberal politicians and their supporters. The Lewinski Presidential sex scandal, Kennedy family drinking problems, and similar shortcomings were a godsend for Limbaugh and he made the most of them.

Limbaugh began his radio career in the 1960s as a teenager in Cape Girardeau, Missouri. The wealth of his locally influential, judge father included owning the radio station Rush began on. He turned professional on Pittsburgh's Top 40 station KQV. Calling himself "Jeff Christie," Limbaugh began to cultivate his radio persona. In the mid-1980s he left radio to work as a promotions director for the Kansas City Royals baseball team, but returned to the airwaves in 1988, first in California then moving on to New York a few years later. It was in New York that Limbaugh's career skyrocketed. He wracked up a huge audience, a hefty salary, fame and stardom, and

Rush Limbaugh talking during his radio program. Copyright Mark Peterson/CORBIS.

impressive credentials by receiving the Marconi Radio Award in 1992, 1995, and 2000, and being inducted into the Broadcast Hall of Fame in 1993.

While the accolades rolled in, Limbaugh's broadcasting became increasingly difficult: he began losing his hearing. He kept this problem a secret as long as he could, but listeners were struck by the changes in his distinctive speech patterns. He and his staff set up a system by which he could appear to be responding directly to their questions. Limbaugh eventually admitted the problem and had a cochlear implant installed in one ear, returning most of his hearing. That tackled, there were other troubles waiting to plague him. Besides what many considered a racist remark on the sports network ESPN about the media intentionally supporting a black NFL quarterback simply because he was black, Limbaugh's personal shortcomings were about to come back to haunt him. Conservative radio talkers took a special joy in uncovering and pointing out the personal flaws of others. They saw things like promiscuity and especially substance abuse as examples of the essential moral weakness of the liberal philosophy. Limbaugh rejected the disease approach to drug addiction, saying users should be shown no mercy and "sent up the river" instead. In 2004 the tabloid *National Inquirer* broke a story that Limbaugh had been addicted for some years to painkillers, particularly one known as Oxycotin. Rumors and accusations about him engaging in the illegal distribution of those drugs flew as Limbaugh checked into a rehabilitation program. Limbaugh's faithful listeners were unfazed while opponents pointed out the central hypocrisy of being guilty of the same things he harangued others for.

SHOCK RADIO

Joining the conservative commentators, religious broadcasters, and sports announcers in the 1980s was a new character, the "shock jock." Where traditional disk jockeys looked for personae that drew attention to themselves, their outrageousness was rather tame and meant to be more cute than off-putting or challenging. Radio station owners watched their on-air personalities carefully to ensure nothing went out over the air that would anger or offend their listeners or advertisers. To help keep any errant words from getting out, stations resorted to a broadcast delay where even though live, the sound came out 5 or 6 seconds later. There was also the "dump button," which when pushed would blank out a section of the broadcast. These technical adaptations helped ensure that nothing unwanted, even the host coughing, would go out over the airwaves. The shock jocks, however, intended to be outrageous and startling, and to break barriers of acceptable radio behavior. The exemplar of this form, and possibly the man for whom the term was coined, was Howard Stern.

Stern was at first compared to Don Imus, much to Stern's chagrin. Born in Arizona in 1940, Imus entered the radio industry in 1971. He was less interested in specific political points of view than he was trying to shock and outrage. He made a number of remarks, thought by some to be anti-Semitic and racist, and suffered through a number of substance-abuse problems. His *Imus in the Morning* show became, in spite of or because of these distinctions, wildly successful. In addition to an enormous audience, he attracted many high-profile guests including the then presidential candidate Bill Clinton in 1992. Because of superficial similarities, Stern and Imus were often linked (the two worked for the same radio station in the early 1980s and were featured together in ads for the station).

What helped Stern thrive in the format, and separated him from Imus and others, was that while working hard at being outrageous, he was a far more complex and subtly subversive character than the simple moniker could describe. He defied classification by being more a political anarchist than either strictly conservative or liberal. His free-form belief system could infuriate those trying to associate him with their causes by one minute espousing conservative causes like gun ownership and in the next breath supporting liberal causes like abortion rights. A self-styled philosopher/sage he could wax philosophic on the career of 1960s radical Abbie Hoffman—arguing that Hoffman was an American hero who saved many young lives (including Stern's own) by helping shorten the war in Vietnam—and then have a stripper as an in-studio guest. No topic was taboo: politics, religion, race, gender, violence, and, of course, sex. He could veer wildly from one point on the political spectrum to another. One day a guest could be a Ku Klux Klan member, the next the social activist Al Sharpton: all of it done with the erratic nature of a horny, know-it-all, fifteen-year-old. He was sharp-witted, politically aware, and more comically original than the host of imitators (particularly the morning "Zoos") who popped up in the wake of his success. He was also an astute businessman who helped expand radio personality syndication. Conventional wisdom argued that such a format would only work locally. Listeners were said to be uninterested in anything that did not pertain to their local condition, especially in the crucial morning drive time slots. Therefore, shows piped in from other geographic locations that did not concern themselves with local weather or traffic conditions would fail. Stern showed this to be a false assumption. He began airing his show, which was based in New York, in Philadelphia and then to an increasing number of other markets across the country, all of which saw their ratings soar.

As the 1980s and 1990s passed, Stern's success grew. He became wealthy, published several books [one a quirky autobiography called *Private Parts* (1993), which was made into a film that was both critically received and a box office success], and branched out into television. Despite the success, Stern paid a price. All the sex and supposed dirty talk ran him straight into the arms of the Federal Communications Commission (FCC). Stern and his radio station owners had been dogged by complaints since the mid-1990s, which found their way to the government. Most of the complaints originated with disgruntled religious group members and obscure morality watchdog organizations.

In 2004 the question of obscenity on the radio rose to new heights. That February the nation's largest radio station chain, Clear Channel Communications, suspended Stern's broadcasts from their outlets. They specifically cited his show of the 24th saying that it was "vulgar." Stern was talking to a man named Rick Salomon who had appeared in a homemade video, engaged in amorous activity with the dubious pop celebrity Paris Hilton, which had recently gained notoriety. Clear Channel president John Hogan was quoted as saying that Stern had been "vulgar, offensive, and insulting, not just to women and African Americans but to anyone with a sense of common decency" (Salant 2004). At the same time, they fired Florida shock jock "Bubba the Love Sponge," real name Todd Clem, after the FCC fined him and Clear Channel $755,000 for lewd on-air remarks. More than a few observers noted that Clear Channel's actions came amidst U.S. congressional hearings on obscenity on the airwaves. Hogan, along with executives from ABC, NBC, and FOX, was scheduled to give testimony. Media watchers pointed out that Stern had been saying such things on his show for years and Clear Channel never seemed to have a problem with it until the heat was turned up by the government. His battles with the government and decency advocates created a peculiar situation.

The Left did not particularly care for Stern's brand of humor. Leftists saw him as a sexist, racist clod opposed to everything they stood for. Yet, he was fighting against the same type of censorship they abhorred. The Right had trouble with Stern for what it considered his prurient preoccupation with sex and anti-Fundamentalist stance. Again, in April of 2004, the FCC fined Clear Channel $495,000 for the obscenity charge. As a result the radio station fired Stern from its network. Hogan called Stern a "liability" and a "risk" they were no longer willing to take. In the midst of the government's sudden burst of energy to stamp out indecency, some argued that the stakes needed to be intensified by raising the amount of the fines imposed. Self-appointed moral critics argued that the FCC's standard fines were so small as to be insufficient threats. Partly as a result the government began to consider the raising of the limit to $500,000 for each violation. It is important to note that the standard procedure of fining radio broadcasters for improper behavior starts at the listener level. A listener had to make the first complaint before the FCC would go into action and tended to confront only those stations specifically mentioned in the complaint. In the Stern case, a complaint was made against a station in Ft. Lauderdale, Florida. The FCC calculated a number by saying that there were three violations during the program in question and fined Clear Channel for six stations, bringing the total to eighteen violations with a separate fine for each.

Stern charged that there was more than just dirty words and suggestive humor involved. He claimed darker motives behind his troubles with the FCC. Stern was openly supportive of President Bush in the first part of his administration. He backed his efforts abroad as well as his domestic programs at home. He faulted Democrats for being weak and fiscally irresponsible. He felt a strong economy key and argued that President Bush was doing a better job of getting the economy going and generating wealth. Throughout the 1990s, Stern generally supported the Republican Party with odes to Governor Pataki and Mayor Giuliani while haranguing former

Democratic New York Mayor David Dinkins and Governor Mario Cuomo. Stern tended to side with the power structure, except of course when it came to the FCC and his personal troubles. In 2004, however, Stern turned his loyalties to the Democratic presidential candidate, Senator John Kerry, and turned on Bush for his sympathy for the religious Right, prosecution of the Iraq war, his opposition to gay marriage, and First Amendment issues. Stern went so far as to ask his listeners to vote against the Republican Party *en masse*. He felt the FCC crackdown was a retaliation for his change of allegiance and then openly anti-Bush position. As a result, in October 2004 Stern announced that he was leaving traditional commercial radio for the new medium of satellite radio: a pay service. He cited government and religious morality group crackdowns as the reason for his move. On such a service there are few government restrictions and virtually none on obscenity and decency. The announcement was seen by radio watchers as something of a sea change for the medium. Just the mention of Stern's intentions sent Sirius Satellite Radio's stock price up and Infinity Broadcasting (Stern's then home) plummeting.

LIBERALS STRIKES BACK

Politically right-wing and conservative radio traditionally centered on individual talk show hosts. In contrast the more left- and liberal radio tradition centered on entire radio stations. Well before the conservative ascendancy, a unique liberal voice began coming out of radio first in California and then in New York. Just after the end of World War II, pacifist Lewis Hill left Washington, D.C. for San Francisco with a dream of creating a radio network that would cater to a more intellectual and socially progressive audience. He wanted it to be more than just a radio station, but also an activist force for social good. In April of 1949 *Pacifica Radio*, as Hill called it, began broadcasting over KPFA-FM in Berkeley. By 1960 a second station, WBAI in New York, had appeared and was tackling the tough social questions of the day. Station hosts and their guests debated marijuana use, nuclear war, the antiwar movements over Korea, Vietnam, and the Gulf, hate speech, civil rights, and many others. *Pacifica* aired some of the first public discussions of the work of literary lights such as Alen Ginsberg and Bertrand Russell. The stations wore their liberalism proudly and railed against anything they saw as undermining the basic liberal values of equality and justice in American and world society. Their reporters covered events from El Salvador to South Africa to the obscure, but volatile island nation of East Timor in the South Pacific. As a "community" radio network *Pacifica* received federal funding, but lived mostly off listener donations. As a result of its liberal stance, conservatives in the government were constantly trying to cut the network's funding. *Pacifica* fought back and won a number of court cases as well as awards for its work, and survived attempts to have it declared indecent.

By the 1990s conservative agendas ruled the AM talk radio world. This challenged the myth of a liberally biased media, though few on the Right were willing to acknowledge it and continued to argue that they were in the minority. In an attempt to counter the ideologically conservative domination of the talk airwaves, *Air America Radio* began broadcasting at the end of March 2004. Spearheaded by comedians Al Franken and Janeane Garofalo, rapper turned activist Chuck D (real name Carlton Douglas Ridenhour), and a host of Liberal leaning comedians and political activists, *Air America* began with start-up funds from Democratic Party supporters and venture capitalists who saw a market for a liberal radio format as well as a way to challenge conservative radio hegemony. Franken argued openly that *Air America's* mission was to directly oppose the conservative establishment and to help remove George Bush from office. Investors leased stations in New York, Chicago, Los Angeles, and San Francisco to form a network (Air America was also the name of the CIA airline that operated in South East Asia during the Vietnam War, performing covert operations against Communist insurgents). Where conservative radio hosts like Limbaugh and Grant tended to the dour and self-referentially important, *Air America* hosts Franken and Garofalo tended to the humorous and self-deprecating as a way to draw listeners into the discussion of weighty topics like politics and war.

Al Franken is best known as one of the original writers and performers on *Saturday Night Live*. A Harvard graduate, Franken's comedy leaned toward political and controversial issues. In the early twenty-first century he became increasingly political in opposition to what he saw as a right-wing takeover of American journalism on television and radio as well as a general drift toward the Right of the U.S. government and the Bush administration in particular. He attacked conservative radio icons like Rush Limbaugh, Bill O'Reilly, and the FOX network. In his books *Rush Limbaugh Is a Big Fat Idiot* (1999) and *Lies and the Lying Liars Who Tell Them* (2003), he poked pointed fun at what he considered the pomposity and hypocrisy of the Right. FOX news unsuccessfully sued Franken over the second book.

The network's other prime comedic host was Janeane Garofalo. Born in the suburbs of New Jersey to conservative parents, Garofalo entered the stand-up comedy world while working on a degree in history at Providence College in the mid-1980s. During this period she won the title of

"Funniest Person in Rhode Island" in a cable television competition. After a chance encounter with actor and filmmaker Ben Stiller in California, her career picked up steam. She received notice for her humor which was of the observational type, but with a sarcastic, intellectual edge. In 1994 she was part of the cast of *Saturday Night Live* and then the innovative liberal political commentary series *TV Nation* created by Michael Moore, who himself had gained notoriety with his caustic documentary film, *Roger and Me* (1989), as well as the highly controversial film critique of the Bush administration, *Fahrenheit 911* (2004). A list of successful film roles followed and she took up permanent residence in lower Manhattan. Throughout this period Garofalo became increasingly active politically, though not necessarily publicly, first in the name of feminist causes, then in opposition to the Clinton administration's activities in the Middle East. After living through the September 11 attacks she was struck with rage at the terrorists, but at the same time anger at the Bush administration. Following the U.S. entry into Afghanistan she became increasingly vocal in her opposition to how the "war on terror" was being waged, and became a more active member of the burgeoning antiwar movement. Because of her public stance, she appeared on a number of political talk shows including Bill O'Reilly's. As a result of her articulate and well-informed opposition to the war she was labeled a traitor and "Saddam Lover." In addition, a television series she had put together for ABC tentatively called *Slice-O-Life* was first accepted and then canceled by the network. ABC executives gave a number of reasons for dropping the project though many media observers cried foul, saying that the network had caved in to opposition to Garofalo for her antiwar activity. In an interview in *Ms.* magazine in the summer of 2003, she argued that "I'm not apologizing, and I'm not letting them shut me up!" (38). With *Air America Radio* Garofalo, Franken, and the others found an eager audience who felt it had been left behind, but now had a voice.

THE LISTENING AUDIENCE

In countries where the government owned and operated the radio system, they regulated what was said and that was that. Political parties and state functionaries gave out only that information, which helped prop up the system, kept people in line, and promoted the approved way of life. In the United States, with its mostly privately owned airwaves, a different situation developed. The battle over censorship in the United States is unlike that going on anywhere else in the world and so is a good subject for the study of the tricky nature of civil rights and freedoms.

With no pictures to deceive the senses or fool the eye, radio forced audiences to concentrate on words and form pictures of their own and to digest the meaning of what was being said. This was a key ingredient in the success of the early dramatic radio shows. The creaking door in the opening of *The Shadow*, for example, conjured up details in the mind a projected image could never create. As a result, radio had to be adept at using words to convey pictures as well as meaning. One of the problems of any spoken language is in its interpretation. A word used in one context could mean one thing, the same word used in a different context could mean something else. How people heard words and interpreted them has always been a major concern for commercial radio broadcasting. Commercial radio's most important constituency was not the listening audience as much as the advertisers. What allowed radio stations to exist was revenue from advertisers who sold products on the airwaves. Visionaries and utopians saw radio's promise as a great democratizer: a medium in which the free exchange of ideas and the coming together of different parts of the human community for the advancement of all. The reality, skeptics argued, was that radio became just another shill for corporate masters whose only concern was the sale of products. The link between radio listening, station popularity, and purchasing power was not lost on either the radio industry or their patrons. In the *2003–2004 Radio Marketing Guide and Fact Book for Audiences* it is made clear that "whether they're in the office, relaxing at home, behind the wheel, or online, your customers have radio as a faithful companion." Advertisers were concerned with what went out over the air where their products were being sold. The idea was to lure listeners in, not drive them away. Little controversy or anything even remotely indecent could be found. Censorship came mostly from advertisers not the government. A corporate sponsor expected advertising agencies to promote its soap, soup, and automobiles in wholesome ways meant to inspire listeners to buy. Radio stations were expected to create a similarly wholesome and nonthreatening platform for the commercials to appear on.

A contributing factor to early radio's nonconfrontational nature was the concept of the "audience." Commercial broadcasting attempted to reach as wide an audience as possible and so took a middle road conceptually. By the 1930s radio programmers and advertising agencies began to realize that there were discrete listening audiences and that going after one built listenership. Stations quickly began following formats. The format idea said that a specific topic was chosen, say country music, then everything about that station would be geared to a country music audience. Disk jockeys were expected to know all the latest country musicians, gossip, and insider information. Advertisers who saw their audiences in the country music world would put their spots there while advertisers who went for more general audiences would produce versions of their commercials with special country music appeal: perhaps getting country music stars to do the voiceover or to use country music as a background. The format dominated what the station's personality was all about and who its audience was. Radio programming leaned more and more toward specific audiences and away from a general audience. When Bruce Morrow, for example, called himself "Cousin," he meant the audience as much as himself. His listeners were his cousins; they were a family linked together through the commonality of the music they liked.

In examining the radio audience there is the concept of the "imagined community." A model for the imagined community of radio is a nation. A "nation" is in many ways an imagined entity where no one member ever knows every other member. At the most, citizens of any given nation know only a fraction of their fellow citizens, yet they still see themselves as a nation. While there are physical borders and confines to any nation the most important link is the ethereal imagined nation: the link of ideas. The nation exists in the mind as much as on the ground. Radio helped form similar communities of like-minded listeners, who at best knew only a few other listeners, yet felt part of the larger community they interacted with through the radio. They shared dreams, values, likes and dislikes and could feel as one. Because most radio was broadcast live, it had a more immediate feeling and effect. The radio puts one in the stands at a ball game, the audience at a concert, at the feet of a political speaker, in the middle of an attack. The community building power of radio was so strong that the mechanical aspect vanished in a conscious way and listeners simply allowed themselves to be drawn in. Connecting with one of these audiences was central to a successful radio station. It grew increasingly apparent that this narrowly focused market approach was not only a viable, but a lucrative venture. What one community or audience might find uninteresting or offensive, another audience would eat up. Bob Grant or Rush Limbaugh's bombastic self-importance or Janeane Garofalo's razor wit and earnest call for humanity mattered little to most corporate advertisers as long as they brought in revenue. As narrow formatting grew in popularity, advertisers grew less interested in the format, only in the size of the audience it brought in. More listeners meant more buyers, regardless if they were conservative or liberal, sports, heavy metal, or rap fans. The requirement of wholesome radio dwindled in the face of profitable radio as advertisers were willing to allow more and more outrageous behavior as long as products were being sold.

Controversial and shock radio showed that these audiences too purchased products. Radio performers had long sought specific audiences for their work. They drew position and power from the number and intensity of their listeners. Bob Grant's various employers were loathe to drop him because he brought in such a huge listening audience. WABC directors estimated that he had an audience of roughly a million. This translated into around $7 million in revenue a year. He appealed to a certain demographic (a specific part of a population) that wanted to hear his message. In 1998 media analyst Herb Norman argued that Grant, and others like him, spoke to a part of the listening community that feared homosexuals were "subverting our youth and threatening the very underpinnings of the nation's morality; that Liberals and fuzzyheaded academics continue to pave the way for a communist takeover." WOR station manager Jerry Crowley said in an interview that Grant had not only faithful listeners, but also faithful advertisers who followed him from one station to another because they knew his listeners did as well. Crowley said that despite the controversy, his station's ratings jumped from 17th to 4th place in the time slot Grant moved into. He also admitted how previously his sales team had difficulty selling space for that time slot, but once Grant was there the advertising almost sold itself. This kind of connection between a radio talk show host and the audience is powerful. They hang on every word, not just the political content but the messages to purchase products as well. Advertisers counted on listeners buying whatever the host sold. This is part of the process by which the host builds his or her audience. In her study of radio audiences, Susan Douglas argued that different eras and generations listened to the radio in such a way that it "shaped the contours of American cultural and political history" (Douglas 1987, 7). How people learned to listen, who they listened to, what types of music or other entertainment, or how they used the experience went a long way to forming their "inner selves" and thus how they interacted with the larger world. The radio was used by individuals and entire communities to change their moods and states of mind, to rally around issues, and to get out votes. It broke down walls between groups in both the public and private worlds. Audiences could also be seen as entire cultures. As cultural influences spread beyond national borders some radio communities tried to protect themselves.

An ethereal medium, radio cannot be stopped by a land border because the wave forms travel regardless of where one country ends and another begins. In the late twentieth century France, for example, underwent anxiety over the dilution of its prized culture. During the nineteenth century, French culture had come to permeate the world and had become the standard for erudite and sophisticated individuals everywhere. By the twentieth century, however, it was U.S. culture that was pushing its way around, and the French culture was being uprooted, even in France itself. This was a common phenomenon in human history. From time to time powerful cultures had worldwide influence: the Egyptians, Greeks, Romans, Chinese, and a host of others had their moment in the sun. Radio allowed this to happen at a greatly enhanced rate. While some welcomed the new culture, others regretted the swamping of their traditional ways and fought back. In France some members of the government supported by like-minded citizens grew resentful of the new cultural inroads as well as the growing number of immigrants from the Middle East and Africa moving into the country. In the mid-1990s, the French music industry was able to convince the Parliament to vote to regulate how much foreign music was played over French radio stations. The new communications rule would mandate that at least 40 percent of the content sent out over French airwaves had to be in French or of French origin (at around the same time the government of Ireland was considering similar legislation). It was an attempt to protect French musical heritage. All of France was seen as one large listening community that wanted to hear French music. Musical quotas were thought a necessary measure to protect a culture in peril. Not everyone in France supported the move, however. Some radio stations argued that they had built their audience upon an international pop format. If they had to drop 40 percent of it, the stations would suffer. The borderless nature of radio called into question the definition of listening community. In the year 2000 Canada was forced to ask similar questions as had the French. With such a powerful radio neighbor as the United States, which community was being served by which culture?

Laura Schlessinger, better known as Dr. Laura, had extensive syndication for her California-based advice talk show, including Canada. She had millions of listeners who tuned in for her strict interpretations of personal behavior and no-nonsense answers and moral absolutes. She was already controversial in the United States for comments she made concerning homosexuality. A trained psychologist, she was called cruel in her responses and took an Old Testament approach to meting out advice saying that the Ten Commandments were the final arbiter of all human behavior. She enraged the gay community with remarks about how gays were "biological errors" and "dysfunctional" people who needed to be converted back to heterosexuality. This outraged many in Canada as well. Her show came to the attention of Ronald Cohen, chairman of the Canadian Broadcast Standards Council (CBSC) when listeners in Toronto and Halifax complained. The CBSC declared that Dr. Laura had been discriminatory and despite her training was not competent to make medical pronouncements about the nature of homosexuality. Cohen asked Schlessinger to publicly apologize to Canadian listeners: she refused. (Like many who delighted in pointing out the shortcomings and moral weakness of others, Dr. Laura had a few of her own. At the height of her fame it was revealed that she had posed for nude pictures some years before.) Canadian radio regulations passed in 1986 prohibited not only the transmission of obscene material, but also any material that was abusive or would expose the target of the speech to hatred or contempt. Not everyone was happy about the CBSC's ruling, including those opposed to Dr. Laura. They feared that the Canadian communication law was too vaguely worded and might be used to limit freedom of speech.

The CBSC was looking to protect Canadian airwaves from what it thought was inappropriate content. In another form of protection, the Canadian government passed the "Hit rule" in 1975. This mandated that FM stations were allowed to play Hit music only 50 percent of the time (a Hit was determined to be music listed in the Top 40). The Hit rule was passed as a way to protect faltering Top 40 AM stations. As a result of better quality sound and an increase in the tendency of FM stations to play more Hit-oriented music, AM stations were losing listeners and disappearing. Implemented by the Canadian Radio–Television and Telecommunications Commission (CRTC) the Hit rule lasted until 1991. Unfortunately, the AM market continued to dwindle and as FM was playing fewer Hits, they too began to disappear from Canadian airwaves. In order to stop this decline, the rule was amended to lower the percentage and exclude any homegrown music from the quota. In addition, French language stations in the bilingual nation argued that English language stations were running them out of business. In 1998 the government cut back the Hit rule even further, allowing Top 40 music to make a comeback in Canadian radio markets. All this underscores the difficulties governments face when trying to regulate radio broadcasting. Many countries still tried to control the medium, especially in the area of obscenity.

OBSCENITY AND CENSORSHIP

As there was little government radio in America, only a few regulatory bodies like the FCC and advocacy groups like the National Association of Broadcasters (NAB), the airwaves were open to a good deal of experimentation. Instead of the government trying to control radio, it was left to big business and self-appointed moral crusaders. Practically speaking, censorship and obscenity meant sex, not politics, or even violence. Bob Grant could call for gays to be executed and G. Gordon Liddy could advise listeners to go for head shots when aiming at federal agents. While they were criticized for their words, it was Howard Stern who was fined and fired for discussing sex. Anything which is meant to create sexual desire is considered pornography. Not all pornography is considered legally obscene and is therefore protected under the First Amendment of the U.S. Constitution, which says:

> Congress shall make no law respecting an establishment of religion, or prohibiting the free exercise thereof; *or abridging the freedom of speech* [my emphasis], or of the press; or the right of the people peaceably to assemble, and to petition the government for a redress of grievances.

There are two types of pornography that are not covered under the First Amendment: obscenity and child pornography. The definition of "obscene" is one of the more problematic of the English language, particularly from a legal standpoint. Even the U.S. Supreme Court has wrestled with it almost to the point of exhaustion. In his famous and oft-quoted line from the 1964 *Jacobellis v. Ohio* case, Justice Potter Stewart said that, while he could not strictly define obscenity, "I know it when I see it." Critics of obscenity law often ask how someone can be kept from doing something or be prosecuted for doing it when no clear definition of the offense exists.

The first local obscenity laws in America date back to the 1830s. The first federal law came in 1842. Following the Civil War, Congress put together a definition of obscenity which included anything that might corrupt a person who might have suicidal tendencies to a breaking point by seeing or hearing it. The Hicklin Rule, as it came to be known, remained until the early 1960s when the Roth-Memoirs test was adopted by the Supreme Court. It had three parts by which material would be judged. Paraphrased, it said the work in question had to appeal to a prurient interest in sex, it had to offend community standards, and it had to be devoid of any redeeming value. The current federal obscenity law was a result of the 1973 *Miller v. California* case. Commonly referred to as the Miller Law, this test is essentially a beefed-up and clarified version of Roth-Memoirs. The FCC explanation is:

> Obscene speech is not protected by the First Amendment and cannot be broadcast at any time. To be obscene, material must meet a three-prong test:
>
> - An average person, applying contemporary community standards, must find that the material, as a whole, appeals to the prurient interest;
> - The material must depict or describe, in a patently offensive way, sexual conduct specifically defined by applicable law; and
> - The material, taken as a whole, must lack serious literary, artistic, political, or scientific value.

Current American radio obscenity laws can be traced back to those first enacted in the 1930s and 1940s. These regulations culminated in 1994 with the implementation of "18 USC 1464" from which the regulations on what could be said originated as well as the penalties (fines and jail time). The problem with such regulations was that no matter how conscientiously lawmakers try to word obscenity laws, the language becomes mired in vague definitions. Also, there was no mention of any material that might not be of a sexual nature, but that still offends. One phrase often employed in the discussion of obscenity on the radio was "community standards." These are supposed to be local ideas about what counts as obscenity. But what is a community? Is it a specific geographic locale, or something less tangible like a larger listening community? If the community which listens to Howard Stern has no problem with what he says, does that protect him from prosecution? Was Stern being obscene or indecent? This is yet another important definition. Talk that is "indecent" falls below that of "obscene" on the rating scale. While closely related to obscene, indecent is not considered as bad an offense and is allowable up to a point as long as it is not engaged in at a time when it is likely that children will be listening.

The definition of indecent was established by the landmark *FCC v. Pacifica* case of 1978. It centered on a monologue performed by comedian George Carlin. An ex-Catholic with a savage sense of humor and astonishing command of the English language, Carlin had developed a joke about things one could not say on radio and television and called them the seven dirty words. Carlin did this routine as part of his stand-up act and it appeared on his album, *Class Clown* (1972). In 1975 the New York radio station WBAI played the recorded version over the air as part of a discussion on obscenity. A listener complained to the FCC about the broadcast because his son had been listening. The FCC approached WBAI for an explanation and then said that if another complaint were received, they would fine and sanction the station. The judgment was overturned, and then overturned again. In its ruling, the Supreme Court recognized what it was doing was to "decide whether the Federal Communications Commission had any power to regulate a radio broadcast that is indecent but not obscene." In a bit of flowery metaphor the court ruled that "when the commission [FCC] finds that a pig has entered the parlor, the exercise of its regulatory power does not depend on proof that the pig is obscene" (no. 77–528). The court was still divided on the issue. Justices Brennan and Marshall dissented with the ruling, saying they were bothered by the increase in what they called the "governmental homogenization of radio." In the end the *Pacifica* dirty words case led to a formalization of indecency regulations in the U.S. broadcast industry. The case led to the creation of the "safe harbor," which allowed indecent radio programming between 10 PM and 6 AM. Lawmakers in the United States, however, seemed to veer back and forth in their rulings on indecency and the radio. In 1997 the U.S. Supreme Court ruled that the Communications Decency Act (CDA) violated the First Amendment and therefore was unconstitutional. The CDA was adopted by Congress the year before as part of the Telecommunications Act of 1996 and was designed to combat child pornography and indecency on the Internet. One last example of the murky difference between sex and violence on the radio is the case of Michael Savage. Less well known nationally than Rush Limbaugh, Savage spewed out a level of right-wing talk astonishing even by conservative radio standards. Based out of San Francisco, his audience was estimated at 7 million and was syndicated on 330 stations mostly in the western part of the country. Despite a protest to stop it, the cable TV news station MSNBC gave Savage a talk show in 2004. After a series of antigay remarks he was fired. He still had a radio job, however, as his bosses were happy with how much revenue he was pulling in. Despite making derogatory remarks over the air about ethnic minorities, Jews, and what he called "gutter slime," he was kept on at KNEW-AM and its affiliates. Ironically, the station owner who refused to drop his show was Clear Channel, the same company who fired Howard Stern for being vulgar.

WHO OWNS RADIO?

As much as what was being said on radio, by the end of the twentieth century, who owned radio was becoming an important issue. Prior to World War II, the only real networks were NBC and CBS in America and the BBC in England. The majority of radio stations in America were independently owned operations with limited range of transmission. Stations were so few and far between in America, for example, that the FCC established what was to be known as the "fairness doctrine." This was an attempt to bring commercial broadcasting to as large an audience as possible. Under the fairness doctrine, stations were encouraged to produce a range of programming. As a result, stations offered a mixed bag of news, music, and sports rather than the highly focused format concept of the late twentieth century. They were thought to have an almost sacred duty to serve the public even though they were private businesses. As a child of the Depression, radio was deemed to have a special place in service to American society just like the railroads, banks, or government. In addition to wide format, the fairness doctrine also said that if programs were aired that were meant for specific audiences, particularly in the realm of politics and religion, the station had to air any opposing views that wished to be heard. As overall radio coverage increased, broadcasters became critical of the fairness doctrine.

To begin with, by the later part of the twentieth century, station owners had recognized the revenue potential of focused formatting. Then station owners feared legal action because they could not possibly give equal time to every group that wanted it—they preferred to give space to those who were willing to pay the highest advertising rates for the privilege of getting airtime. In 1985, after a series of legal challenges, the Supreme Court ruled that radio stations were not required to give or sell time to opposing groups in order to meet the requirements of the fairness doctrine as long as the station continued to meet the ill-defined "public trust." Proponents of the fairness doctrine saw this ruling as another salvo in the conservative agenda to eliminate multiple points of view in the realm of public political radio discourse. The FCC report of the same year went a step further by questioning the doctrine of fairness itself. It said that a station following the fairness doctrine would be inhibited from airing anything controversial lest it be forced to air an opposing view. In 1987, the FCC tried to force a Syracuse, New York, television station to air opposing views after it aired spots urging listeners to support the local construction of a nuclear power plant. The ensuing court case prompted the FCC to finally repeal the fairness doctrine. The repeal of the fairness doctrine contributed to the rise of conservative talk radio as stations could then not just follow a specific music format, but an overtly political one as well without the need to air opposing views. This also made it easier for station owners to link stations together and put out specific agendas.

Another FCC ownership regulation was the "seven station rule," which limited the number of radio stations a single owner could have. This was meant, like the fairness doctrine, to promote diversity among the airwaves and prevent monopolies forming in the radio industry the way they had in railroads, oil, and steel. A number of radio–robber barons opposed this, claiming that the FCC did not have the authority to regulate the formation of networks. In the *NBC v. The United States* case, the Supreme Court ruled that the powers Congress gave the FCC did allow them to regulate networks even though their charter did not specifically say so. The so-called "network case" was subsequently used by the FCC to support its regulation of network alliances and ownership quotas. By the 1980s, however, this changed. Under President Ronald Reagan, with his predilection for deregulation, radio broadcasters were given increasing latitude in dealing with the FCC and the commission itself was reigned in by the Republican-controlled government. With the soaring number of licenses being granted (as well as considerable pressure being applied by the networks and owners on Congress and the FCC), rules pertaining to how many radio stations could by owned be an individual person or corporation were all but eliminated.

There are those who argue that when large corporations own many radio stations, especially ones in the same area, the public's access to unbiased information is curtailed. What alarms some is the labyrinth of corporate ownership which can be hard to untangle and can lead to interesting occurrences. In New York, for example, the major AM talk radio station, WABC, is owned by Capitol Cities Communications, which in turn is owned by the Disney Corporation. According to Frank Rich, a respected *New York Times* reporter, when Bob Grant was fired from WABC the order came from Disney's upper management. Though Rich disliked Grant, he was bothered by the apparent arbitrary nature of the firing and at how corporate types would hire and fire on a whim if they felt revenue was at stake. To further complicate the matter, syndicated radio attorney Alan Dershowitz was also fired by WABC for calling Grant a racist. Dershowitz, however, was hired by WOR. Then the National Association of Radio Talk Show Hosts nominated Grant, Dershowitz, and Michael Eisner for "Freedom of Speech Awards" (Eisner was the chairman of Disney who was thought to have been responsible for Grant's firing).

Possibly the thing that worries many radio listeners and First Amendment advocates more than dirty words on radio is the concept of consolidation. Begun in the mid-1990s, radio consolidation was a deregulation policy that made it easier for large corporations to buy up as many radio stations as they wanted and consolidate them into meganetworks. Opponents of consolidation argued that it would allow a few powerful corporations to own the majority, or even all, of the stations. They also argued that as the big radio conglomerates expanded, listener choice would shrink. What is known as radio consolidation began when President Bill Clinton signed the Telecommunications Act of 1996. It was meant primarily to cover the deregulation of telephone and television service, but included a section that raised the number of radio stations a single owner could have in one market and removed all restriction on nationwide ownership. The act thus allowed for any single person or corporation to own as many radio stations across the country as they wanted. The big corporations fought hard to get this clause included and immediately took advantage of it. Led by companies such as Westinghouse, Chancellor Media, and Clear Channel, those who had the financial resources to do so began gobbling up smaller local stations. The big companies, supported vigorously by the NAB, argued that consolidation would benefit the radio industry by bringing greater resources to small stations, making radio more widely available, and greatly diversifying radio content for larger audience.

Opponents of consolidation saw darker implications in a few large owners controlling so much, if not all of radio. They pointed out that large corporations tend to dislike any critiques of their practices or anything that might hamper sales. As such, critics argued, the big corporate radio networks would not air any news or other information that would contradict the corporate outlook and as such news would become sanitized, music bland, and social and political activism nonexistent. Opponents of consolidation pointed out that in order to meet corporate revenue quotas, radio stations began playing more commercials and less content (polls showed that 30 percent of listeners complained about too many commercials per hour). Workaday radio people also suffered. Radio analyst Todd Spencer said that from 1996 to 2002 as many as 10,000 radio jobs were lost due to consolidation. Critics argued that consolidation was leading to a homogenization of the radio where all the stations played the same music over and over, the deejays sounded the same, and that this was killing the fragile existence of radio "community." Antibig radio forces argued that the diversity the corporations point to was artificial. They say the same genres the NAB said showed diversity—contemporary hit, pop, adult urban and others—are carefully constructed by the corporate networks to promote nonthreatening blandness meant to create an atmosphere suitable for advertisers. (Listening to any of the morning "Zoos" that seem to cover the country from one coast to another, with their calculated "wackiness" and "irreverence," one would find it hard to argue against the creeping blandness.) Critics also argue that consolidation is helping eliminate places where young musicians, commentators, and radio hosts can go to get their work heard. If every station just plays Rush Limbaugh or Howard Stern, where can new hosts get work? If all the music that is played is Britney Spears, how will audiences ever get to hear the next Bruce Springsteen or Tupac Shakur? How will concerned citizens know what is going on around the country or around the world in times of crisis, or hear varied and independent political ideas when the corporate interests give only a narrow view?

Despite the ravages of consolidation, small radio may still revive. In 2004 Congress began working to allow for low-power noncommercial FM stations to begin broadcasting. The idea of small-scale radio stations was first put forward in 1999 to counter consolidation and the trend toward bigger and bigger corporate radio giants. The big radio groups opposed the low-power concept, arguing that those stations would cause reception interference and static for the mainstream outlets. Low-power advocates charged that the big guys were using interference as a ruse to keep them off the air and out of competition. Low-power supporters found a sympathetic ear when the FCC held hearings on the issue. The NAB argued that the FCC's study, known as the Mitre Report, and which showed that the small stations would not cause static interference to the larger stations, was flawed. After hearing the NAB's report, Congressional panel chairman Senator John McCain shot back referring to the NAB, in a moment of pique, as being owned by Clear Channel Communications. McCain supported the low-power station concept as a way of bringing diversity back into radio (part of the deal would be that the low-power stations could not be bought and sold or consolidated into larger entities). As a result, radio will continue its evolution.

CONCLUSION

By the early twenty-first century, radio had come a long way from *Amos 'n' Andy*. Radio around the world was then at the center of a storm of questions about weighty subjects like rights and freedoms, the responsibility of the state to the individual and the individual to the state: questions yet to be answered. Radio had reached a point where genuine interaction between broadcaster and listener was occurring. The private messages between amateur enthusiasts in the 1920s had given way to interactive communication on a mass scale with people discussing and arguing crucial issues. The combination of radio and the telephone allowed the once one-way conversation to run both directions. Even the least controversial talk or music show now includes listener participation. There are dark clouds, however, but also bright sunshine in this situation. It is still too soon to know if it will be the utopians or the robber barons who win out.

7

Conclusion: Did Video Kill the Radio Star?

With the advent of the Internet, radio was able to add a new dimension to itself. It was discovered that radio signals could be digitized and transmitted over phone lines and, thus, over the Internet and into computers anywhere on earth. In the 1990s many radio stations added their transmissions to the World Wide Web. This allowed for a cheap and easy way for "local" stations to be heard worldwide: something only the British Broadcasting Corporation (BBC) came close to doing in the past. Unlike the British effort, the Web allowed even small radio stations to reach a potentially huge listening audience. The other major advance in radio technology at the opening of the twenty-first century is satellite radio. This is a commercial free, pay service that listeners subscribe to, like cable or satellite television. Listeners choose the content they receive depending upon their musical likes or dislikes—talk radio and weather are also available. The satellite concept began in the late 1990s when the Federal Communications Commission (FCC) allotted an "S" band of 2.3 GHz for broadcasting. A communications satellite was placed into a stationary, or geosynchronous, orbit over the United States at a height of around 30,000 miles. A ground station beams a signal to the satellite, which then transmits it to the subscriber's radio receiver. Part of the advantage of satellite radio is that wherever the subscriber goes, the music follows; they no longer lose the station if they travel far away from it. This system could prove to be a major shift in radio listening.

Along with satellite there was the appearance of stations with limited transmission range but wide-ranging ideas about broadcasting. These "IndieRadio" stations have made a move back to the earliest days of wireless transmitting when hosts played the music they wanted and discussed the topics they wanted, and transmitted it to audiences even more narrowly focused than usual. These mostly music oriented–outlets substituted polished deejays in favor of enthusiastic amateurs, whose lack of professionalism was made up for by sheer passion for music. Also known as NeoRadio, these seemingly small and independent stations are mostly controlled by powerful corporate giants like Clear Channel. Part of the reason for the megaradio involvement in the NeoRadio movement is because of the latest advance in radio technology: high definition radio (HDR). The big radio group's plan is to be part of the new technology instead of being swamped under by it. HDR greatly improves sound quality, making songs played on a car radio sound as if the listener was using their car CD player. HDR also allows broadcasters to run several shows at the same time on the same station, so listeners could go to that station and listen to their preferred content without abandoning the station altogether. Ultimately, such a service as HDR, like high definition television services, would allow the subscriber to choose their own music: in essence to be one's own deejay.

There are also the looming phenomena of pod casting. The home computing technology of the late twentieth century, which allowed users to download music in the form of MP3 files, allowed enterprising users to put their favorite music online and swap them with other Web surfers for free. Online sites such as Napster became enormously popular and resulted in government and music industry crackdowns. Alternative pay–MP3 services soon appeared. In 2005 a number of individuals, including former MTV host Adam Curry, began digitizing and making available not just music, but their deejay voice-over to go with it. A service subscriber could now download a radio show and listen to it at their convenience. Such a concept would allow anyone without their own radio station to broadcast a show of their own by simply recording it as an MP3 file and putting it online for downloading.

Religious leaders, inventors, business people, and politicians have all tried, with varying levels of success, to control "The Media" and its electronic forms of communication. Just as Napoleon knew in the early nineteenth century that those who control the means of communication can control the empire, modern day dictators have tried to do the same. In the early twenty-first century, it is common to hear complaints and dire warnings about the corporate takeover of radio and other forms of communication. While these groups certainly try to exert control, there has been a backlash.

Fiercely independent phenomenon like *Pacifica Radio*, *Air America*, the low-power FM movement, and even the whacky fringe political world with its ranting have helped keep some corporate or government colossus from seizing total control. It shows that if enough segments of society resist it complete control cannot be achieved. The radio may very well fulfill its utopian role of protector of democracy and the free exchange of ideas.

Arguments were made that the radio was helping spread vulgar pop culture that undermined the society. As were culture critics ranting about what the child radio was putting out, so was daddy. Lee De Forest, whose invention of the audion—a vacuum tube that allowed for the appearance of modern radio—was not shy about calling himself the father of radio. As a disappointed parent he was also not shy about complaining that his little baby had gone afoul of the wrong crowd and was dashing all the dreams he had for his progeny. He railed against the base new pop culture that was taking over; he had seen radio as an uplifting mode of civilization. Classical music and other highbrow entertainment would show people a better way. Others, however, would argue that De Forest's child had done just that. Millions around the world had been uplifted by what they heard on the radio. They saw a future there, brighter than the one they were living and it inspired them.

The involvement of radio in the 2004 American presidential election helped show that perception of voices without pictures could go off in many directions. Despite the saturation of television coverage, radio still played a crucial role. The hearing extension of radio allowed the listener to become a participant in the political process although at first only listening. The rush of enthusiasm felt by the resurgence of liberal radio ran head-on into the harsh reality of the reelection of George Bush. The Right saw media like radio as a major part of getting out the vote and bringing it a mandate to rule. Undaunted, the Left vowed to fight on, particularly in making accusations of voter fraud across the country and in efforts to point out what they saw as the inequities and misdirection of the ruling elite and to rally listeners to civic action. The conservative talkers lost no time in boasting of "their" victory and taking credit for it. One of the more important side effects of the hotly contested election was the appearance of the Blog: an ingenious marriage of the Internet and the radio. Blogging is a more immediate, visceral, and public form of e-mail that allowed listeners to comment in real time on the live discussion on the air and act as a new grassroots form of journalism. A number of media commentators argued that the sudden growth of the Blog was a direct result of a growing disillusionment and resentment over the perception that the mainstream media had given over its duty to be watchdogs of the government and corporate world, and had joined in with them. Bloggers, often ordinary concerned citizens and would-be muckrakers shut out of the mainstream, took matters into their own hands and in essence created an alternate source of news that was not hooked to corporate power, but was more democratic. The liberal media outlets, *Air America* in particular, but conservative stations as well, made extensive use of the Blog concept and viewed it as a fundamental reconfiguration of radio news and political commentary. Bloggers argued that they were doing something that the mainstream news media was not: properly informing the electorate of the important core issues and investigating political fraud and vice, instead of the fluffy nonsense and irrelevancies the networks were increasingly focusing on. The marriage of radio and Internet may prove to be a new era in electronic communication, which will spread beyond the United States and inject radio with a new life, with citizens becoming more active in the running of government and other pressing issues of the day.

This book has been a biography of a technology. Radio began in a swirl of enthusiasm for a new medium which it was thought could make the world a better place and which could be engaged in by anyone. Throughout the twentieth century, efforts were made to eliminate that freewheeling and democratic nature and to position it firmly under government or corporate control. The early twenty-first century saw what may be the seeds of a rebirth of that original grass roots spirit. This period may not be radio's old age, but rather its genuine maturing or even rebirth. Much has been made of the power of such technologies as genetic engineering, robotics, and nanotechnology to alter not just our society, but humans ourselves. While radio cannot change one's genetic makeup, give one an immortal mechanical body, or send armies of little machines into one's bloodstream to repair damaged tissue, it can change one's mind. Those who might have thought "video killed the radio star" had to take a second look at the squawking sound box and realize that what was thought to be a doddering old codger could still kick like a youth. The ability to get a person to think in a new way or become politically active or simply make you feel better because you heard your favorite song—something radio has always been able to do—is a powerful technology indeed.

Glossary

AC (Alternating Current). An electrical current where the electrons can reverse their direction of travel.

Account Executive. Station salesperson who arranges for advertising.

Actives. Listeners who participate in the life of the station by calling in to make requests and comments.

Ad lib. An unrehearsed and spontaneous improvisational comment.

AM (Amplitude Modulation). A system for transmitting audio by changing the strength of a radio wave.

Amplification. Increasing the power of a signal.

AOR. Album-Oriented Rock radio format sometimes called Classic Rock.

AP. Associated Press, a syndicated news service.

ASCAP. American Society of Composers, Authors, and Publishers.

Audio. Any form of recorded sound.

Audion. Lee De Forest's improved Fleming valve.

Bandwidth. The range of frequencies of radio waves put out by a transmitting station.

Bed. A piece of instrumental music used to back up speech in radio programs.

Blasting. Excessive volume which creates distortion.

BMI. Broadcast Music Incorporated.

Call letters. The numbers and letters used to identify a station to listeners.

Carrier wave. The signal which radiates from the transmitter and carries the actual sound.

CART (from cartridge). A means of storing audio clips either on special tapes or in a digital system.

Channel. A section of the audio mix machine where all the controls relate to a single source of audio.

Commercial. An on-air advertisement.

Conductor. A material which allows the passage of electrical current.

Consultant. Advisor or counselor to the radio station on various issues.

Control room. Center of broadcast operations from which programming originates.

Co-op. An arrangement between retailer and manufacturer for the purpose of sharing radio advertising expenses.

Copy. Text of an advertising message or performance script.

CPB. Corporation for Public Broadcasting.

D.A.B. (Digital Audio Broadcasting). A system for transmitting audio by sending a set of digital signals.

D.A.T. (Digital Audio Tape). A high-quality audio recording system.

DC (Direct Current). An electrical current that runs in one direction only.

Dead air. Silence where sound usually should be, or an unintentional absence of programming.

Decibel (dB). A unit for measuring the loudness of an audio signal.

Deejay (from disk jockey). The host of radio music program who plays music.

Demographics. Statistical data pertaining to the age, sex, race, income of the listening audience.

Drive time. Radio's prime time normally from 6–10 AM and 3–7 PM, when many listeners are doing so from their cars.

Electromagnet. A magnet created by electricity flowing through a coil of wire.

Fade. To slowly lower or raise volume level.

FCC. Federal Communications Commission.

Feature. A live or recorded element of a radio program.

Feedback. A form of distortion created if a speaker is on while a nearby microphone is.

Fidelity. Trueness of sound dissemination or reproduction.

Fleming valve. The early form of vacuum tube.

FM (Frequency Modulation). A system for transmitting audio by varying the frequency of a radio wave. Produces higher quality sound than AM.

Format. Type of programming a station offers.

Frequency. Number of cycles per second of a sine wave.

Gain. A method of adjusting sound levels. Similar to volume control.

Hertz (Hz). Cycles per second; unit of electromagnetic frequency.

Heterodyning. When two or more radio signals overlap causing distortion.

Hype. Exaggerated presentation; high-intensity, punched.

ID. Station identification using call letters required by law.

Insulator. A material which prevents the passage of electric current.

Jingle. Music, commercial, or promotion used as an audio logo for a station.

Kilohertz (kHz). One thousand cycles per second.

Level. Amount of volume units.

Licensee. An individual or a company which holds the license issued by the FCC for broadcast purposes.

Link. Speech that connects two recorded radio broadcast items together.

Market. Area served by a broadcast facility.

Megahertz (MHz). Million cycles per second; FM frequency measurement, megacycles.

Mono. Single sound; monaural, monophonic as opposed to stereo.

MOR. Middle-of-the-Road radio format.

Multiplex. A bundle of digital radio channels on a single frequency.

Multiplexing. Impressing two or more signals on one carrier as in FM stereo.

NAB. National Association of Broadcasters.

Narrowcasting. Directed programming which targets a specific audience demographic.

Network. Broadcast combine which includes a number of local stations into one corporate entity.

Network feed. Programs sent to affiliate stations.

Ohm. A unit of measure of the resistance of the flow of electrons over a wire.

Output. Transmission of audio or power.

PACKAGE. A short recorded feature containing edited extracts from interviews, vox-pops, scripted links and other material. Usually produced by a reporter for a magazine or news programme.

Passives. Listeners who do not call stations but only listen in.

Payola. Illegal payment to a disc jockey or radio station programmer for playing a particular piece of music.

Playlist. List of music for airing.

Plug. Promotion for a client's product.

Production. The process of preparing audio for playback.

Promo (from promotional). An advertisement for a product or the station itself.

PSA (Public Service Announcement). An advertisement for a nonprofit charitable cause.

Radio frequency. The number of times the radio wave vibrates per second.

Rating. Measurement of the total available audience.

Reach. Measurement of how many different members of an audience will be exposed to a message.

Regeneration. Technique of feeding back a signal inside a vacuum tube to increase its power.

Remote. Broadcast originating outside of the radio station, normally at the site of some event.

Reverb. An echo.

Segue. Uninterrupted flow of recorded material; continuous.

Share. Percentage of station's listenership compared to that of competing stations.

Signal. Sound transmission.

Simulcast. Simultaneous broadcast over two or more frequencies.

Sound bite. A small portion of a larger speech or interview.

Spectrum. Range of frequencies available to broadcasters.

Sponsor. A station's advertiser or client.

Spots. Commercials.

Station. Broadcast facility given a specific frequency by the FCC.

Station log. Document containing specific day-to-day operations of a station.

Stereo. Sound coming from both the right and the left loudspeakers.

Super heterodyne. The modern vacuum tube which allowed for the operation of radio and television.

Syndication. Programs sent to a network of users.

Talent. Radio performers of any kind.

Toll broadcasting. Precursor to commercial radio.

Trailer. Used to promote forthcoming programs.

Transmit. To broadcast.

Tuning. Adjusting a radio receiver to pick up a particular frequency.

Underwriter. Program sponsor.

Voice-over. Talk over sound.

Watt. A unit of measure for electrical power.

Wireless telegraphy. Device used to transmit Morse code but not sound without the need for transmission wires.

WRAP. A short feature, where a reporter has wrapped their voice around an interview or music.

Bibliography

Adair, Gene. *Thomas Alva Edison: Inventing the Electric Age.* New York: Oxford University Press, 1996.

Aitken, Hugh G. J. *The Continuous Wave.* Princeton, NJ: Princeton University Press, 1985.

Anderson, Benedict. *Imagined Communities: Reflections on the Origin and Spread of Nationalism.* London and New York: Verso, 1991.

Archer, Gleason. *History of Radio to 1926.* New York: American Historical Society, 1938.

Auletta, Ken. *Three Blind Mice: How the TV Networks Lost Their Way.* New York: Random House, 1991.

Benjamin, Louise M. "In Search of the Sarnoff 'Radio Music Box' Memo." *Journal of Broadcasting and Electronic Media* 37 (1993): 325–335.

Bick, Jane Horowitz. "The Development of Two-Way Talk Radio in America." Diss., University of Massachusetts, 1987.

Cantor, Geoffrey, David Gooding, and Frank A.J.L. James. *Michael Faraday.* Atlantic Highlands, NJ: Humanities Press, 1996.

Cardwell, Donald. *Wheels, Clocks, and Rockets: A History of Technology.* New York: W.W. Norton & Co., 1995.

Cheney, Margaret. *Tesla: Man Out of Time.* New York: Dell Publishing, 1983.

Cohen, I. Bernard. *Science and the Founding Fathers: Science in the Political Thought of Jefferson, Franklin, Adams, and Madison.* New York: W.W. Norton & Co., 1995.

Cowan, Ruth Schwartz. *A Social History of American Technology*. New York: Oxford University Press, 1997.

Crocq, Phillippe, and Emmanuel Legrand. "French Radio Required to Meet Repertoire Quotas." *Billboard*, January 8, 1994.

De Forest, Lee. *Father of Radio: The Autobiography of Lee De Forest*. Chicago: Wilcox & Follett, 1950.

Dibner, Bern. *Early Electrical Machines*. Norwalk, CT: Burndy Library, 1957.

———. *Ten Founding Fathers of the Electrical Science*. Norwalk, CT: Burndy Engineering Co., 1954.

Douglas, Susan J. *Inventing American Broadcasting 1899–1922*. Baltimore, MD: Johns Hopkins University Press, 1987.

———. *Listening In: Radio and the American Imagination*. New York: Random House, 1999.

Dreher, Carl. *David Sarnoff: An American Success Story*. New York: Quadrangle/New York Times Book Co., 1977.

FCC v. Pacifica Foundation, No. 77-528, US Supreme Court ruling, 438US726. Argued April 18–19, 1978. Decided July 3, 1978.

Fischer, Claude. *America Calling: A Social History of the Telephone to 1940*. Berkeley: University of California Press, 1992.

Geselowitz, Michael N. "The Life of Hugo Gernsback, or Why Is There No Such Thing as 'Engineering Fiction?'" *Today's Engineer* (April–May, 2002): Webzine.

Grace, Kevin Michael. *When In Doubt, Censor*, Alberta Report, June 5, 2000.

Greene, Stephen L. W. "Who Said Lee de Forest Was the 'Father of Radio?'" *Mass Communication Review* (1991).

Hangen, Tona. *Redeeming the Dial: Radio, Religion, and Popular Culture in America*. Chapel Hill: University of North Carolina Press, 2002.

Haring, Kristen. "The 'Freer Men' of Ham Radio: How a Technical Hobby Provided Social and Spatial Distance." *Technology and Culture* 44 (2003): 734–761.

Harlow, Alvin. *Old Wires and New Waves: The History of the Telegraph, Telephone and Wireless*. New York: D. Appleton-Century Co., 1936.

Hawkes, Ellen. "Shock and Jaw: Climate of a New Blacklist?" *Ms.* XIII:2 (Summer 2003): 31-39.

Hilmes, Michele. *Radio Voices: American Broadcasting, 1922–1952*. Minneapolis: University of Minnesota Press, 1997.

Holland, Bill. "Support Grows For Low-Power FMs." *Billboard*, March 6, 2004.

Isaacson, Walter. *Benjamin Franklin: An American Life*. New York: Simon & Schuster, 2003.

Jackson, Camille. *The Rating Game*, Southern Poverty Law Center Intelligence Report, Spring, 2004.

Jacot de Boinod, Bernard. *Marconi—Master of Space: An Authorized Biography of the Marchese Marconi*. London: Hutchinson & Co., 1935.

Lasar, Matthew. *Pacifica Radio: The Rise of an Alternative Network*. Philadelphia, PA: Temple University Press, 2000.

Leblanc, Larry. "FM 'Hit' Rules Arose for AM Support." *Billboard*, November 28, 1998.

Leblanc, Larry, and Sean Ross. "Canadian Top 40 Format Enjoying a Resurgence." *Billboard*, February 14, 1998.

Leinwoll, Stanley. *From Spark to Satellite: A History of Radio Communications*. New York: Charles Scribner's Sons, 1979.

Levinson, Paul. *The Soft Edge: A Natural History and Future of the Information Revolution*. New York: Routledge, 1997.

Lewis, Tom. *Empire of the Air: The Men Who Made Radio*. New York: E. Burlingame Books, 1991.

Lucas, Allison. "Selling a Controversial Character." *Sales and Marketing Management*, July, 1996.

Lynch, Don, and Ken Marschall. *Titanic: An Illustrated History*. New York: Hyperion, 1992.

Maclaurin, W. Rupert. *The Clark Collection: Invention and Innovation in the Radio Industry*. New York: Macmillan Co., 1949.

Marconi, Degna. *My Father*. Ottawa: Balmuir Book Pub., 1982.

Marsh, Dave. *Born to Run: The Bruce Springsteen Story*. New York: Doubleday and Co., 1979.

Morse, Samuel F. B. *An Argument on the Ethical Position of Slavery in the Social System, and Its Relation to the Politics of the Day*. Society for the diffusion of political knowledge. New York: 1863.

———. *The Foreign Conspiracy against the United States*. New York: Leavitt, Lord, and Co., 1835.

Morus, Iwan Rhys. *Frankenstein's Children: Electricity, Exhibition, and Experiment in Early-nineteenth-century London*. Princeton, NJ: Princeton University Press, 1998.

Munson, Wayne. *All Talk: The Talkshow in Media Culture*. Philadelphia, PA: Temple University Press, 1993.

Nobel Lectures, Chemistry 1922–1941. Amsterdam: Elsevier Publishing Co., 1966.

Nobel Lectures, Physics 1901–1921. Amsterdam: Elsevier Publishing Co., 1967.

Norman, Herb. "The Talk of the Town: The Saga of Bob Grant." *Journal of Popular Culture* 32 (1998): 91.

Patnode, Randall. "What These People Need Is Radio: New Technology, the Press, and Otherness in 1920s America." *Technology and Culture* 44 (2003): 285–305.

Pera, Marcello. *The Ambiguous Frog: The Galvani-Volta Controversy on Animal Electricity*. Translated by Jonathan Mandelbaum. Princeton, NJ: Princeton University Press, 1992.

Persico, Joseph. *Edward R. Murrow: An American Original*. New York: McGraw-Hill, 1988.

Raby, Ormond. *Radio's First Voice: The Story of Reginald Fessenden.* Toronto: Macmillan of Canada, 1970.

Radio Marketing Guide and Fact Book for Advertisers: 2003–2004. New York: Radio Advertising Bureau, 2003.

Salant, Jonathan D. "Shock Jock Howard Stern Pulled off the Air," *AP release*, February 25, 2004.

Schiffer, Michael. *The Portable Radio in American Life.* Tucson: University of Arizona Press, 1991.

Segrave, Kerry. *Payola in the Music Industry: A History 1880–1991.* Jefferson, NC: McFarland and Co., 1994.

Selligmann, Jean, and Carol Hill. "Rave on, Tuned Out." *Newsweek*, April 29, 1996.

Shields, Todd, and Katy Bachman. "Media Wire." *Media Week*, January 19, 2004.

Spangler, Matt. "Can't Find Nothin' On Radio?" *Radio and Records*, July 31, 1998.

Spencer, Todd. "Radio Killed the Radio Star." *Salon.com*, October 1, 2002.

Standage, Tom. *The Victorian Internet.* New York: Walker and Co., 1998.

Sterling, Christopher and John Kittross. *Stay Tuned: A Concise History of American Broadcasting.* 2nd ed. Belmont, CA: Wadsworth Publishing Co., 1990.

Taylor, Church. "Weiland's Excesses Get STP Music Iced; the Bob Grant Saga Continues at WOR." *Billboard*, May 11, 1996.

Wasserman, Neil. *From Invention to Innovation: Long Distance Telephone Communication at the Turn of the Century.* Baltimore, MD: Johns Hopkins University Press, 1985.

White, Thomas H. "United States Early Radio History." http://www.earlyradio history.us.

Whittaker, Edmund. *A History of the Theories of Aether and Electricity.* Vol. 1, *The Classical Theories.* Los Angeles, CA: Tomash Publishers, 1987.

Index

About the Author

BRIAN REGAL teaches American history and the history of science and technology at the TCI College of Technology in New York—the school originally founded by Guglielmo Marconi in 1909. His previous publications include *Henry Fairfield Osborn: Race and the Search for the Origins of Man* (2002) and *Human Evolution: A Guide to the Debates* (2004). His most recent article is "Maxwell Perkins Editor of Eugenics" in *The Princeton University Library Chronicle* (February 2005).